Genetic Catastrophe!
Sneaking Doomsday?

Genetic Catastrophe! Sneaking Doomsday?

with
A Dictionary of Genetic Damage

by Nils K. Oeijord

Writers Club Press
San Jose New York Lincoln Shanghai

Genetic Catastrophe! Sneaking Doomsday?
with
A Dictionary of Genetic Damage

Writers Club Press
an imprint of iUniverse, Inc.

For information address:
iUniverse, Inc.
5220 S. 16th St., Suite 200
Lincoln, NE 68512
www.iuniverse.com

ISBN: 0-595-22565-9

Printed in the United States of America

To the gene-damaged victims of science (via pollution, including radiation), technology (via pollution, including radiation), and other human activities such as, for example, smoking and diagnostic doses of X-rays.

Contents

Introduction

"We [the human race] do not have much time to prove that we are not the product of a lethal mutation."

Science 263: 181, 1994

"I almost think it is the ultimate destiny of science to exterminate the human race."

Thomas Love Peacock

This book uses American examples of gene/genetic damage to the human species. However, the human genes are as much damaged in the rest of the world as they are in the US. This book is written to defend, not to attack, the US, and the rest of the world. The genetic catastrophe consists of four major genetic "epidemics"—those of cancer, vascular disease, musculoskeletal disease, and behavioral disease. There are identified approximately 250 common genetic diseases, and approximately 7,000 "rare" genetic diseases. However, the real numbers are much larger, and, remember, this is only the beginning. The list of genetic damage is growing daily. The natural rate at which mutations and genetic damage occur is reasonably constant. The larger the gene the more common genetic diseases of the gene are. Hence, more common genetic diseases are caused by mutations/damage to very large genes. Workshops are developing guidelines for management of children with several common genetic diseases. Guidelines for several "rare" genetic diseases are currently being prepared. Genetic damage is far more common than generally understood. There is insufficient recognition of the magnitude of the genetic catastrophe. The concept of

1

genetic disease has expanded during the last twenty years. The concept has now expanded to be virtually all encompassing. Even infectious diseases have some relationships to our genes. Noble-laureate Paul Berg said "all human disease is genetic."

A species is the genes' way of making more genes. The ability to select helpful genes, and to evolve, is absolutely necessary for the survival of a species, including the human species. The *natural* mutation frequency is 1 mutation per 100,000 genes per generation. This value proves that present-day pollution is an absolute genetic tragedy because this value lies *near* the critical value that separates evolution from extinction.

The Genetic Catastrophe!
Evolution Derailed?

Our genes and chromosomes have been damaged, with tragic consequences. The strong increase in the gene damage of genetic diseases like cancer, vascular diseases, Alzheimer's, Parkinson's, diabetes, osteoporosis, obesity, etc, is perhaps the best illustration of the magnitude of the genetic catastrophe. The genetic catastrophe is an unintended consequence of some well-intentioned effort to improve human life by sidestepping nature. Certainly the victims of genetic damage are the victims of modern human activities (including science and science based technology). Scientists, governments, bureaucracies, politicians, and the rest of us are asleep at the wheel. So who will save us?

About 85 percent of us are tortured and finally killed by cancer, vascular diseases, and other non-rare genetic diseases. Most of the rest of the population are tortured, and often finally killed, by so-called rare diseases.

Science, in general, does not speak about a genetic catastrophe. On the contrary, by using words like syndrome, disorder, disease, illness, defect, deficiency, failure, etc, instead of *genetic damage,* science, in general, is defining away and covering up the genetic catastrophe. The dictionary below tries to counter this policy.

In the US, a rare disease is defined as one that afflicts no more than 200,000 people. However, one out of 10 Americans suffers from just such a rare disease. And more than 6,000 diseases have been classified as rare, according to the US National Organization for Rare Disorders. If (when) these 6,000 diseases reach, on average, the 200,000 people level, then each American, on average, will suffer from more than four rare diseases. Most rare diseases are genetic diseases caused by gene damage. In our globally polluted world our genes suffer damage just through day-to-day living. Our DNA is constantly under attack from mutagenic chemicals and radiation causing a steady build up of heritable gene damage, in spite of the best efforts of our DNA repair enzymes and other repair mechanisms, such as the SOS repair mechanism, the cell-suicide mechanism, sterility, infertility, natural and unnatural abortion, recurrent miscarriages, death at an early age, and natural and unnatural selection.

There is no evidence that there is a dose below which there is not a mutagenic effect. A small dose of chemicals/radiation to a large population does more harm to the gene pool than a large dose to a small population. In fact, the effluents causing the gene damage currently meet all environmental standards, and often the gene damage-causing chemicals are present at non-detectible levels. We don't know what's actually happening at the genetic level. Besides, genetic damage is particularly insidious because it can take several generations for the effects to show up. Moreover, the total exposure to mutagenic chemicals and radiation is unknown. Genetic damage is the most important, but most neglected, issue of our time.

The blood of Inuit people and polar bears contains a large number of mutagenic chemicals. Certainly, the romantic notions of "wilderness", "health", "science", and "technology" are outdated. Although the individual is often powerless to avoid exposure to widely used chemicals, there are examples of mutagens to which people voluntarily expose themselves, for example cigarette smoke. Unfortunately, mutagenic chemicals occur in widely divergent chemical groups, ranging from simple compounds such as formaldehyde to complex ones such as alkaloids. With every breath of cigarette smoke, the body is confronted by more than 800 mutagenic chemicals which include dioxin-like compounds. And apropos of dioxins, the genetic damage of dioxins is absolutely devilish: The DNA double-helix acts like a zipper, which can open and close. A dioxin molecule acts like a thread-like object which is preventing the zipper from being closed!!! All mother's milk (human and nonhuman) of the world contains dioxins, now and in the future.

Some 150,000 (an increasing number) American infants are born annually with birth defects, which include brain abnormalities and cleft palate. These 150,000 infants are probably born in our average environment. "44 million Americans can't read" because of the genetic damage of dyslexia. An individual may be superintelligent but he/she is unable to learn to read because of a specific genetic damage. These 44 million (an increasing number) individuals are probably born in, and live in, our average environment. The genetic damage of cancer kills approximately 30 percent of the total US population, while the genetic damage of cardiovascular diseases kills approximately 40 percent of the US population. A 20-year-old individual may suddenly die of a cardio-vascular disease.

Because every human individual has about 35,000 functional genes, there is an endless number of possible heritable genetic diseases. Furthermore, there is an endless number of possible chromosome anomalies such as chromosome breaks. Ionizing radiation is extremely efficient at causing chromosome breaks. Even gasoline vapor and high-voltage fields cause chromosome breaks. These messy breaks are difficult or impossible for cells to repair correctly. Evidence shows that the cell's repair-system is fallible even when it is confronted only by a minimal challenge. Obviously, certain occupations cause certain gene/chromosome damage. Even relatively low (diagnostic) doses of X-rays cause gene damage. Data strongly suggest that preconceptional exposure of the mother to diagnostic doses of X-rays increases the risk of offspring with Down's syndrome. Chemicals in the general pollution damage crucial DNA repair genes. Besides, heavy metals damage repair enzymes (change their form and function). When a repair-gene is damaged, the gene damage will magnify the consequences of the cell's subsequent exposures to all mutagens (chemicals and radiation), because of the cell's diminished ability to repair gene damage correctly. And free radicals and other mutagens attack our DNA all the time. Cells somehow sense gene damage and activate a suicide program, called apoptosis, to kill themselves so that genetic damage is not perpetuated. This is the ultimate "brake" against cancer as well. The p53-gene is known to trigger apoptosis in response to genetic damage in a cell. If the p53-gene is damaged, then an important way of preventing genetic damage from being passed on to the next generation is lost. How can we avoid a genetic damage explosion? Our children will face an uncertain future where the rules of the game are dictated by genetic diseases, behavioral and physical. They will face a world where no place is safe, not even the sanctity of your own soul.

Evolution is a fact, not a theory. Genetic diseases appear to be identical across species. Is our DNA, and thereby our evolution, derailed? Is humanity derailed?

The elimination of genetic diseases in animals and plants can only be accomplished through selective breeding. This is not a theory, this is a fact. However, as we have shown above, the situation for our genes is enormously difficult. It looks as if the genetic diseases will be the winner of the race. The international society should, as soon as possible, prepare for a tribunal, or at least a truth commission, to deal with the worst crime against humanity.

Today twice as many die of diabetes (per 100,000 people) as before insulin was introduced. Today, in general, twice as many die of a genetic disease (per 100,000 people) as before a medicine was introduced. Why? The theory (the fact) of natural/unnatural selection gives the answer. The genetic damage to human sperm cells is so enormous that we can recognize directly, by looking through a light microscope, that about 50 percent of the cells are abnormal. There is perhaps nothing that can prevent all sperm cells to become abnormal. After all, normal cells live in the same polluted environment as the abnormal cells. Of course, the genetic damage to egg cells is even more dramatic because they are relatively few in number. Men born with a birth defect have a doubled risk, compared with other fathers, of having a child with a birth defect, a large population study revealed. Note that this situation, according to simple mathematics, leads to a genetic damage explosion (in a non-evolutionary society). The genetic catastrophe of breast cancer develops extremely fast: Of all the women with breast cancer, only a tenth have family histories of the disease, and half of this group has a heritable genetic damage causing breast cancer! We are now discovering the price of medicine, or, more correctly, the price of a non-evolutionary, or anti-evolutionary, society. The question is now whether the price is worth paying if we look at the *total* impact of medicine on our species. (The immediate impact of medicine on *individuals* is a totally different question.)

Department of Education estimates that 20 percent of Americans are learning-disabled. This result is obviously wrong, because the dyslexics (20 percent of Americans) are not the only learning-disabled people in the US. In the general population in Rochester, N.Y., 3 percent had (in the year 2000) the genetic damage of Tourette's syndrome, and 20 percent had the (genetic?) damage of a tic disorder. The rate of 3 percent in the general population is about 50 to 75 times higher than typical estimates. The number of 6,000 for rare diseases in the US is also wrong. Behind each of the 6,000 diseases (mostly genetic diseases) there are lots of different types of gene damage. America, and the rest of the world, do suffer a silent genetic catastrophe. Science doesn't recognize the genetic catastrophe. Why? The answer is obvious.

It has been estimated that more than 50,000 chemicals are in common use in the United States. Most of these chemicals have not been tested for mutagenicity. Of those that have been tested for mutagenicity, about 20 percent produced mutations in the Ames test. Of the several million chemicals in "uncommon" use in the world, perhaps as much as 1,000,000 are causing gene damage (mutations). All of the "uncommon" chemicals are working in your cells at this very moment. Make no mistake about it: unless we wake up and act courageously, our species will be wiped out. Anyone who truly understands this, must shoulder the immense responsibility associated with this impending genetic catastrophe.

The natural (evolutionary) mutation rate is less than one mutation per fertilized egg. It has been estimated that one out of five persons carries a spontaneous (natural) mutation not present in his or her parents. If the mutation rate exceeds one mutation per fertilized egg, then evolution is derailed. Also, if the mutation rate is one, or slightly less than one, mutation per fertilized egg, *and*, if, at the same time, natural selection is disturbed (it is), then evolution is derailed. And if evolution is derailed, then a genetic catastophe is unavoidable. We are rapidly approaching the point of no return. Humanity is at stake.

Behavior is a biochemical event. Behavioral disorders often have a genetic basis. Most neurobehavioral syndromes are genetic damage. With regard to survival of the human species, damage to genes for behavioral traits ("behavioral genes") are much more dangerous than damage to genes for physical traits. A large percentage of the population suffers from genetic damage of behavioral disorders (and/or mental retardation) such as learning disabilities, depression, antisocial personality disorder, Tourette's syndrome, schizophrenia, bipolar disorder, eating disorders, attention deficit disorder, hyperactivity, fragile-X syndrome, autism, Lesch-Nyhan syndrome, Down syndrome, etc. If (or when) only 100 genetic behavioral disorders reach (on average) the 1 percent level, then humanity is destroyed? No doubt about it: We must preserve our genes and our evolution now. Tomorrow, it will be too late.

Readers should use the content of the following dictionary as general background information representing the enormity of the genetic catastrophe. Note that there are two types of genetic diseases: *inherited genetic diseases*, and *acquired genetic diseases*. The entries of the dictionary do not follow a standard format, and the list does not provide information on synonyms.

A Dictionary of Genetic Damage

0–9

11-Beta-hydroxylase deficiency, genetic damage of

17-Beta-hydroxysteroid dehydrogenase deficiency, genetic damage of

18-Hydroxylase deficiency, genetic damage of

2-Hydroxyglutaricaciduria, genetic damage of

3 Hydroxyisobutyric aciduria, genetic damage of

3 Methylglutaconyl-CoA hydratase deficiency, genetic damage of

3-Hydroxyacyl-CoA dehydrogenase, long chain, deficiency of, genetic damage of

3-Methylcrotonylglycinuria, genetic damage of

3C syndrome, genetic damage of

3M syndrome, genetic damage of

4-Alpha-hydroxyphenylpyruvate hydroxylase deficiency, genetic damage of

4-Hydroxyphenylacetic aciduria, genetic damage of

46 XX Gonadal dysgenesis epibulbar dermoid, genetic damage of

47 XXX syndrome, genetic damage of

47 XXY syndrome, genetic damage of

47 XYY syndrome, genetic damage of

48 XXXY syndrome, genetic damage of

48 XXYY syndrome, genetic damage of

49 XXXXX syndrome, genetic damage of

49 XXXXY syndrome, genetic damage of

5-Nucleotidase syndrome, genetc damage of

5-Alpha-reductase 2 deficiency, genetic damage of
5-Oxoprolinase deficiency, genetic damage of
6-Pyruvoyl-tetrahydropterin synthase deficiency, genetic damage of
7-Dehydrocholesterol reductase deficiency, genetic damage of

A

Aagenaes syndrome, genetic damage of
Aarskog Ose Pande syndrome, genetic damage of
Aase Smith syndrome, genetic damage of
Aase syndrome, genetic damage of
ABCD syndrome, genetic damage of
Abdallat Davis Farrage syndrome, genetic damage of
Abdominal aortic aneurysm, genetic damage of
Abdominal cystic lymphangioma, genetic damage of
Abdominal defects, genetic damage of
Abdominal musculature, absent microphthalmia joint laxity, genetic damage of
Abdominal neoplasms, genetic damage of
Aberrant subclavian artery, genetic damage of
Ablepharon macrostomia syndrome, genetic damage of
Ablutophobia, genetic damage of (possible)
Abnormal systemic venous return, genetic damage of
Abruzzo Erickson syndrome, genetic damage of
Absent corpus callosum cataract immunodeficiency, genetic damage of
Absent T lymphocytes, genetic damage of
Acalvaria, genetic damage of
Acanthocytosis, genetic damage of
Acanthocytosis chorea, genetic damage of
Acanthosis nigricans muscle cramps acral enlargement, genetic damage of
Acarophobia, genetic damage of (possible)
Acatalasemia, genetic damage of
Accessory pancreas, genetic damage of
Acetyl-CoA alpha-glucosaminide-N-acetyl transferase deficiency, genetic damage of
Achalasia, genetic damage of
Achalasia alacrimia syndrome, genetic damage of

Achalasia microcephaly, genetic damage of
Achalasia, familial esophageal, genetic damage of
Achalasia-Addisonianism-Alacrimia syndrome, genetic damage of
Achard syndrome, genetic damage of
Achard-Thiers syndrome, genetic damage of
Acheiropodia, genetic damage of
Achondrogenesis, genetic damage of
Achondrogenesis Kozlowski type, genetic damage of
Achondrogenesis type 1A, genetic damage of
Achondrogenesis type 1B, genetic damage of
Achondrogenesis type 2, genetic damage of
Achondroplasia, genetic damage of
Achondroplasia Swiss type agammaglobulinemia, genetic damage of
Achondroplastic dwarfism, genetic damage of
Achromatopsia, genetic damage of
Achromatopsia incomplete, X-linked, genetic damage of
Acid maltase deficiency, genetic damage of
Acidemia, isovaleric, genetic damage of
Acidemia, propionic, genetic damage of
Ackerman syndrome, genetic damage of
Acne rosacea, genetic damage of
Acoustic neurofibromatosis, genetic damage of
Acoustic neuroma, genetic damage of
Acoustic schwannomas, genetic damage of
Acousticophobia, genetic damage of (possible)
Acquired autoimmune hemolytic anemia, genetic damage of (possible)
Acquired hypoprothrombinemia, genetic damage of (possible)
Acquired ichthyosis, genetic damage of (possible (possible)
Acquired prothrombin deficiency, genetic damage of (possible)
Acral dysostosis dyserythropoiesis, genetic damage of
Acral renal mandibular syndrome, genetic damage of

Acro coxo mesomelic dysplasia, genetic damage of
Acro fronto facio nasal dysostosis, genetic damage of
Acrocallosal syndrome, Schinzel type, genetic damage of
Acrocephalopolydactyly, genetic damage of
Acrocephalosyndactyly Jackson Weiss type, genetic damage of
Acrocephaly pulmonary stenosis mental retardation, genetic damage of
Acrocraniofacial dysostosis, genetic damage of
Acrodermatitis, genetic damage of (possible)
Acrodermatitis enteropathica, genetic damage of
Acrodysostosis, genetic damage of
Acrodysplasia scoliosis, genetic damage of
Acrofacial dysostosis ambiguous genitalia, genetic damage of
Acrofacial dysostosis atypical postaxial, genetic damage of
Acrofacial dysostosis Catania form, genetic damage of
Acrofacial dysostosis Preis type, genetic damage of
Acrofacial dysostosis Rodriguez type, genetic damage of
Acrofacial dysostosis Weyers type, genetic damage of
Acrofacial dysostosis, Nager type, genetic damage of
Acrofacial dysostosis, Palagonia type, genetic damage of
Acrokeratoelastoidosis of Costa, genetic damage of
Acromegaloid changes cutis verticis gyrata corneal, genetic damage of
Acromegaloid facial appearance syndrome, genetic damage of
Acromegaloid hypertrichosis syndrome, genetic damage of
Acromegaly, genetic damage of
Acromesomelic dysplasia, genetic damage of
Acromesomelic dysplasia Brahimi Bacha type, genetic damage of
Acromesomelic dysplasia Campailla Martinelli type, genetic damage of
Acromesomelic dysplasia Hunter Thompson type, genetic damage of
Acromesomelic dysplasia, Maroteaux type, genetic damage of

Acromicric dysplasia, genetic damage of
Acroosteolysis dominant type, genetic damage of
Acroosteolysis neurogenic, genetic damage of
Acroosteolysis osteoporosis skull and mandible changes, genetic damage of
Acropectorenal field defect, genetic damage of
Acropectorovertebral dysplasia, genetic damage of
Acrophobia, genetic damage of (possible)
Acropigmentation of Dohi, genetic damage of
Acrorenal field defect ectodermal dysplasia diabetes, genetic damage of
Acrorenal syndrome recessive, genetic damage of
Acrorenoocular syndrome, genetic damage of
Acrospiroma, genetic damage of
ACTH deficiency, genetic damage of
ACTH resistance, genetic damage of
Activated protein C resistance, genetic damage of
Acutane embryopathy, genetic damage of
Acute articular rheumatism, genetic damage of
Acute erythroblastic leukemia, genetic damage of
Acute febrile neutrophilic dermatosis, genetic damage of
Acute idiopathic polyneuritis, genetic damage of
Acute intermittent porphyria, genetic damage of
Acute lymphoblastic leukemia, genetic damage of
Acute lymphoblastic leukemia congenital sporadic aniridia, genetic damage of
Acute lymphocytic leukemia, genetic damage of
Acute megakaryoblastic leukemia, genetic damage of
Acute monoblastic leukemia, genetic damage of
Acute myeloblastic leukemia type 1, genetic damage of
Acute myeloblastic leukemia type 2, genetic damage of
Acute myeloblastic leukemia type 3, genetic damage of
Acute myeloblastic leukemia type 4, genetic damage of

Acute myeloblastic leukemia type 5, genetic damage of
Acute myeloblastic leukemia type 6, genetic damage of
Acute myeloblastic leukemia type 7, genetic damage of
Acute myeloblastic leukemia with maturation, genetic damage of
Acute myeloblastic leukemia without maturation, genetic damage of
Acute myelocytic leukemia, genetic damage of
Acute myelogenous leukemia, genetic damage of
Acute myeloid leukemia (generic term), genetic damage of
Acute myeloid leukemia, secondary, genetic damage of
Acute myelomonocytic leukemia, genetic damage of
Acute non-lymphoblastic leukemia (generic term), genetic damage of
Acute posterior multifocal placoid pigment epitheliopathy, genetic damage of
Acute promyelocytic leukemia, genetic damage of
Acute renal failure, genetic damage of
Acyl-CoA dehydrogenase, medium chain, deficiency of, genetic damage of
Acyl-CoA dehydrogenase, short chain, deficiency of, genetic damage of
Acyl-CoA dehydrogenase, very long chain, deficiency of, genetic damage of
Acyl-CoA oxidase deficiency, genetic damage of
Adactylia unilateral dominant, genetic damage of
Adam complex familial, genetic damage of
Adams Nance syndrome, genetic damage of
Adams-Oliver syndrome, genetic damage of
Addison's disease, genetic damage of
Adducted thumb syndrome recessive form, genetic damage of
Adducted thumbs Dundar type, genetic damage of
Adenine phosphoribosyltransferase deficiency, genetic damage of
Adenocarcinoid tumor, genetic damage of

Adenocarcinoma, genetic damage of
Adenoid cystic carcinoma, genetic damage of
Adenoma, genetic damage of
Adenoma of the adrenal gland, genetic damage of
Adenomelablastoma, genetic damage of
Adenomyosis, genetic damage of
Adenosine deaminase deficiency, genetic damage of
Adenosine monophosphate deaminase deficiency, genetic damage of
Adenosine triphosphatase deficiency, anemia due to, genetic damage of
Adenylosuccinate lyase deficiency, genetic damage of
Adie syndrome, genetic damage of
Adiposa dolorosa, genetic damage of
Adolescent benign focal crisis, genetic damage of
Adrenal adenoma, familial, genetic damage of
Adrenal cancer, genetic damage of
Adrenal disorder, genetic damage of
Adrenal gland hyperfunction, genetic damage of
Adrenal gland hypofunction, genetic damage of
Adrenal hyperplasia, congenital, genetic damage of
Adrenal hypertension, genetic damage of
Adrenal hypoplasia congenital, X linked, genetic damage of
Adrenal incidentaloma, genetic damage of
Adrenal insufficiency, genetic damage of
Adrenal macropolyadenomatosis, genetic damage of
Adrenal medulla neoplasm, genetic damage of
Adrenocortical carcinoma, genetic damage of
Adrenogenital syndrome, genetic damage of
Adrenoleukodystrophy, genetic damage of
Adrenoleukodystrophy, autosomal, neonatal form, genetic damage of
Adrenoleukodystrophy, X linked, genetic damage of

Adrenomyeloneuropathy, genetic damage of
Adrenomyodystrophy, genetic damage of
Adult onset Still's disease, genetic damage of
Adult spinal muscular atrophy, genetic damage of
Adult syndrome, genetic damage of
Aerophobia, genetic damage of (possible)
Afibrinogenemia, genetic damage of
Aganglionosis, total intestinal, genetic damage of
Aganthia holoprosencephaly situs inversus, genetic damage of
Aggressive fibromatosis, genetic damage of
Aglossia adactylia, genetic damage of
Agnosia, primary visual, genetic damage of
Agonadism dextrocardia diaphragmatic hernia, genetic damage of
Agonadism mental retardation delayed bone age, genetic damage of
Agoraphobia, genetic damage of (possible)
Agrizoophobia, genetic damage of (possible)
Agyria pachygyria polymicrogyria, genetic damage of
Agyria-pachygyria type 1, genetic damage of
Agyrophobia, genetic damage of (possible)
AHD, genetic damage of
Ahumada-Del Castillo syndrome, genetic damage of
Aicardi syndrome, genetic damage of
Aicardi-Goutieres syndrome, genetic damage of
Aichmophobia, genetic damage of (possible)
Ailurophobia, genetic damage of (possible)
Akaba Hayasaka syndrome, genetic damage of
Akesson syndrome, genetic damage of
Aksu Stckhausen syndrome, genetic damage of
Al Awadi Teebi Farag syndrome, genetic damage of
Al Frayh Facharzt Haque syndrome, genetic damage of
Al Gazali Al Talabani syndrome, genetic damage of
Al Gazali Aziz Salem syndrome, genetic damage of

Al Gazali Donnai Muller syndrome, genetic damage of
Al Gazali Hirschsprung syndrome, genetic damage of
Al Gazali Khidr Prem Chandran syndrome, genetic damage of
Al Gazali Sabrinathan Nair syndrome, genetic damage of
Alagille-Watson syndrome, genetic damage of
Alar nasal cartilages coloboma of telecanthus, genetic damage of
Albers-Schonberg disease, genetic damage of
Albinism, genetic damage of
Albinism deafness syndrome, genetic damage of
Albinism immunodeficiency, genetic damage of
Albinism ocular, genetic damage of
Albinism ocular late onset sensorineural deafness, genetic damage of
Albinism oculocutaneous, Hermansky-Pudlak type, genetic damage of
Albinism, yellow mutant type, genetic damage of
Albinoidism, genetic damage of
Albrecht Schneider Belmont syndrome, genetic damage of
Albright like syndrome, genetic damage of
Albright Turner Morgani syndrome, genetic damage of
Albright's hereditary osteodystrophy, genetic damage of
Albuminurophobia, genetic damage of (possible)
Aldolase A deficiency, genetic damage of
Alektorophobia, genetic damage of (possible)
Aleukemic leukemia cutis, genetic damage of
Alexander's disease, genetic damage of
Alkaptonuria, genetic damage of
Allain Babin Demarquez syndrome, genetic damage of
Allan Herndon syndrome, genetic damage of
Allanson Pantzar McLeod syndrome, genetic damage of
Allergic angiitis, genetic damage of
Allergic autoimmune thyroiditis, genetic damage of
Allergic encephalomyelitis, genetic damage of

Allgrove syndrome, genetic damage of
Alliumphobia, genetic damage of (possible)
Allodoxaphobia, genetic damage of (possible)
Aloi Tomasini Isaia syndrome, genetic damage of
Alopecia, genetic damage of
Alopecia anosmia deafness hypogonadism syndrome, genetic damage of
Alopecia areata, genetic damage of
Alopecia congenita keratosis palmoplantaris, genetic damage of
Alopecia contractures dwarfism mental retardation, genetic damage of
Alopecia epilepsy oligophrenia syndrome of Moynaha, genetic damage of
Alopecia epilepsy pyorrhea mental subnormality, genetic damage of
Alopecia hypogonadism extrapyramidal disorder, genetic damage of
Alopecia immunodeficiency, genetic damage of
Alopecia macular degeneration growth retardation, genetic damage of
Alopecia mental retardation hypogonadism, genetic damage of
Alopecia mental retardation syndrome, genetic damage of
Alopecia totalis, genetic damage of
Alopecia universalis, genetic damage of
Alopecia universalis onychodystrophy vitiligo, genetic damage of
Alpers disease, genetic damage of
Alpha 1-antitrypsin deficiency, genetic damage of
Alpha-2 deficient collagen disease, genetic damage of
Alpha-ketoglutarate dehydrogenase deficiency, genetic damage of
Alpha-L-fucosidase deficiency, genetic damage of
Alpha-L-iduronidase deficiency, genetic damage of
Alpha-mannosidosis, genetic damage of
Alpha-sarcoglycanopathy, genetic damage of

Alpha-thalassemia, genetic damage of
Alpha-thalassemia-abnormal morphogenesis, genetic damage of
Alport syndrome, genetic damage of
Alport syndrome macrothrombocytopenia, genetic damage of
Alport syndrome, dominant type, genetic damage of
Alport syndrome, recessive type, genetic damage of
Alstrom's syndrome, genetic damage of
Alternating hemiplegia, genetic damage of
Alternating hemiplegia of childhood, genetic damage of
Alveolar hypoventilation syndrome, genetic damage of
Alveolar soft part sarcoma, genetic damage of
Alveolitis, extrinsic allergic, genetic damage of
Alves Dos Santos Castello syndrome, genetic damage of
Alzheimer disease, familial, genetic damage of
Alzheimer's disease, genetic damage of
Amathophobia, genetic damage of (possible)
Amaurosis congenita of Leber, genetic damage of
Amaurosis congenita of Leber, type 1, genetic damage of
Amaurosis congenita of Leber, type 2, genetic damage of
Amaurosis hypertrichosis, genetic damage of
Amaxophobia, genetic damage of (possible)
Ambral syndrome, genetic damage of
Ambras syndrome, genetic damage of
Ambulophobia, genetic damage of (possible)
AMC, genetic damage of
Amegakaryocytic thrombocytopenia, genetic damage of
Amelia cleft lip palate hydrocephalus iris coloboma, genetic damage of
Amelia facial dysmorphism, genetic damage of
Amelia X linked, genetic damage of
Amelo cerebro hypohidrotic syndrome, genetic damage of
Amelogenesis imperfecta, genetic damage of
Amelogenesis imperfecta local hypoplastic form, genetic damage of

Amelogenesis imperfecta nephrocalcinosis, genetic damage of
Ameloonychohypohidrotic syndrome, genetic damage of
Amenorrhea, primary, genetic damage of
Aminoacidopathies, genetic damage of
Aminoaciduria, genetic damage of
Aminopterin like syndrome without aminopterin, genetic damage of
Amniotic bands, genetic damage of
Ampola syndrome, genetic damage of
Amychophobia, genetic damage of (possible)
Amylo-1,6-glucosidase deficiency, genetic damage of
Amyloid angiopathy, genetic damage of
Amyloid polyneuropathy, transthyretin related, genetic damage of
Amyloidosis, genetic damage of
Amyloidosis of gingiva and conjunctiva mental retardation, genetic damage of
Amylopectinosis, genetic damage of
Amyoplasia, genetic damage of
Amyoplasia mandibulofacial dysostosis, genetic damage of
Amyotonia congenita, genetic damage of
Amyotrophic lateral sclerosis, genetic damage of
Amyotrophy fat tissue anomaly, genetic damage of
Anablephobia, genetic damage of (possible)
Anaphylaxis, genetic damage of
Anaplastic large cell lymphoma, genetic damage of
Anaplastic thyroid cancer, genetic damage of
Andermann syndrome, genetic damage of
Andersen's disease, genetic damage of
Andre syndrome, genetic damage of
Androgen insensitivity syndrome, genetic damage of
Androphobia, genetic damage of (possible)
Anemia, genetic damage of
Anemia sideroblastic spinocerebellar ataxia, genetic damage of

Anemia, pernicious, genetic damage of
Anemia, sideroblastic, genetic damage of
Anemophobia, genetic damage of (possible)
Anencephalus, genetic damage of
Anencephaly, genetic damage of
Anencephaly spina bifida X linked, genetic damage of
Aneurysm, genetic damage of
Aneurysm of sinus of Valsalva, genetic damage of
Angel shaped phalango epiphyseal dysplasia, genetic damage of
Angelman syndrome, genetic damage of
Angiofollicular ganglionic hyperplasia, genetic damage of
Angiofollicular lymph hyperplasia, genetic damage of
Angioimmunoblastic with dysproteinemia lymphadenopathy, genetic damage of
Angiokeratoma mental retardation coarse face, genetic damage of
Angioma hereditary neurocutaneous, genetic damage of
Angioneurotic edema hereditary due to C1 esterase deficiency, genetic damage of
Angiosarcoma of the liver, genetic damage of
Angiosarcoma of the scalp, genetic damage of
Angiotensin renin aldosterone hypertension, genetic damage of
Aniridia absent patella, genetic damage of
Aniridia ataxia renal agenesis psychomotor retardation, genetic damage of
Aniridia cerebellar ataxia mental deficiency, genetic damage of
Aniridia mental retardation syndrome, genetic damage of
Aniridia ptosis mental retardation obesity familial, genetic damage of
Aniridia renal agenesis psychomotor retardation, genetic damage of
Aniridia type 2, genetic damage of
Aniridia, sporadic, genetic damage of
Ankle defects short stature, genetic damage of

Ankyloblepharon cleft palate ectodermal defects, genetic damage of

Ankyloblepharon ectodermal defects cleft lip palate, genetic damage of

Ankyloblepharon filiforme adnatum cleft palate, genetic damage of

Ankyloblepharon filiforme imperforate anus, genetic damage of

Ankyloglossia heterochromia clasped thumbs, genetic damage of

Ankylosing spondylarthritis, genetic damage of

Ankylosing spondylitis, genetic damage of

Ankylosing vertebral hyperostosis with tylosis, genetic damage of

Ankylosis of teeth, genetic damage of

Annular constricting bands, genetic damage of

Annular pancreas, genetic damage of

Annuloaortic ectasia, genetic damage of

Ano-rectal atresia, genetic damage of

Anodontia, genetic damage of

Anonychia ectrodactyly, genetic damage of

Anonychia microcephaly, genetic damage of

Anonychia onychodystrophy, genetic damage of

Anonychia onychodystrophy brachydactyly type B, genetic damage of

Anophthalia pulmonary hypoplasia, genetic damage of

Anophthalmia cleft lip palate hypothalamic disorder, genetic damage of

Anophthalmia cleft palate micrognathia, genetic damage of

Anophthalmia esophageal atresia cryptorchidism, genetic damage of

Anophthalmia megalocornea cardiopathy skeletal anomalies, genetic damage of

Anophthalmia microcephaly hypogonadism, genetic damage of

Anophthalmia plus syndrome, genetic damage of

Anophthalmia short stature obesity, genetic damage of

Anophthalmia Waardenburg syndrome, genetic damage of

Anophthalmos with limb anomalies, genetic damage of
Anophthalmos, clinical, genetic damage of
Anorchia, genetic damage of
Anorchidism, genetic damage of
Anorectal anomalies, genetic damage of
Anorexia nervosa, genetic damage of
Anosmia, genetic damage of
Anotia, genetic damage of
Anotia facial palsy cardiac defect, genetic damage of
Ansell Bywaters Elderking syndrome, genetic damage of
Anterior horn disease, genetic damage of
Anterior pituitary insufficiency, familial, genetic damage of
Anthophobia, genetic damage of (possible)
Anti-factor VIII autoimmunization, genetic damage of
Anti-HLA hyperimmunization, genetic damage of
Anti-plasmin deficiency, congenital, genetic damage of
Antigen-peptide-transporter 2 deficiency, genetic damage of
Antinolo Nieto Borrego syndrome, genetic damage of
Antiphospholipid syndrome, genetic damage of
Antisocial personality disorder, genetic damage of
Antisynthetase syndrome, genetic damage of
Antithrombin deficiency, congenital, genetic damage of
Antley-Bixler syndrome, genetic damage of
Antlophobia, genetic damage of (possible)
Anyane Yeboa syndrome, genetic damage of
Aorta-pulmonary artery fistula, genetic damage of
Aortic aneurysm, genetic damage of
Aortic arch anomaly peculiar facies mental retardation, genetic damage of
Aortic arch interruption, genetic damage of
Aortic arches defect, genetic damage of
Aortic coarctation, genetic damage of
Aortic dissection lentiginosis, genetic damage of

Aortic supravalvular stenosis, genetic damage of
Aortic valve stenosis, genetic damage of
Aortic valves stenosis of the child, genetic damage of
Aortic window, genetic damage of
APECED syndrome, genetic damage of
Apert like polydactyly syndrome, genetic damage of
Apert syndrome, genetic damage of
Aphalangia hemivertebrae, genetic damage of
Aphalangia syndactyly microcephaly, genetic damage of
Aphthous stomatitis, genetic damage of
Apiphobia, genetic damage of (possible)
Aplasia cutis autosomal recessive, genetic damage of
Aplasia cutis cleft palate epidermolysis, genetic damage of
Aplasia cutis congenita, genetic damage of
Aplasia cutis congenita dominant, genetic damage of
Aplasia cutis congenita epibulbar dermoids, genetic damage of
Aplasia cutis congenita intestinal lymphangiectasia, genetic damage of
Aplasia cutis congenita of limbs recessive, genetic damage of
Aplasia cutis congenita recessive, genetic damage of
Aplasia cutis myopia, genetic damage of
Aplastic anemia, genetic damage of
Apo A-I deficiency, genetic damage of
Apolipoprotein C-II deficiency, genetic damage of
Apparent mineralocorticoid excess, genetic damage of
Apple peel syndrome, genetic damage of
Apraxia, buccofacial, genetic damage of
Apraxia, classic, genetic damage of
Apraxia, constructional, genetic damage of
Apraxia, ideational, genetic damage of
Apraxia, ideokinetic, genetic damage of
Apraxia, ideomotor, genetic damage of
Apraxia, motor, genetic damage of

Apraxia, oculomotor, genetic damage of
Apudoma, genetic damage of
Aqueductal stenosis, X linked, genetic damage of
Arachnodactyly ataxia cataract aminoaciduria mental retardation, genetic damage of
Arachnodactyly mental retardation dysmorphism, genetic damage of
Arachnoid cysts, genetic damage of
Arachnoiditis, genetic damage of
Arakawa'sa syndrome II, genetic damage of
Arc syndrome, genetic damage of
Aredyld syndrome, genetic damage of
Arginase deficiency, genetic damage of
Argininosuccinate synthetase deficiency, genetic damage of
Argininosuccinic aciduria, genetic damage of
Arhinia choanal atresia microphthalmia, genetic damage of
Arithmophobia, genetic damage of (possible)
Arnold Stckler Bourne syndrome, genetic damage of
Arnold-Chiari malformation, genetic damage of
Arnold-Chiari syndrome, genetic damage of
Aromatase deficiency, genetic damage of
Aromatic l amino acid decarboxylase deficiency, genetic damage of
Aromatic L-amino acid decarboxylase deficiency, genetic damage of
Arrhinia, genetic damage of
Arrhythmogenic right ventricular dysplasia, genetic damage of
Arroyo Garcia Cimadevilla syndrome, genetic damage of
Arrythmogenic right ventricular dysplasia, familial, genetic damage of
Arterial dysplasia, genetic damage of
Arterial tortuosity, genetic damage of
Arteriovenous malformation, genetic damage of
Arthritis, genetic damage of

Arthritis short stature deafness, genetic damage of
Arthritis, juvenile, genetic damage of
Arthrogryposis, genetic damage of
Arthrogryposis due to muscular dystrophy, genetic damage of
Arthrogryposis ectodermal dysplasia other anomalies, genetic damage of
Arthrogryposis epileptic seizures migrational brain disorder, genetic damage of
Arthrogryposis IUGR thoracic dystrophy, genetic damage of
Arthrogryposis like disorder, genetic damage of
Arthrogryposis like hand anomaly sensorineural, genetic damage of
Arthrogryposis multiplex congenita, genetic damage of
Arthrogryposis multiplex congenita CNS calcification, genetic damage of
Arthrogryposis multiplex congenita distal, genetic damage of
Arthrogryposis multiplex congenita distal type 1, genetic damage of
Arthrogryposis multiplex congenita distal type 2, genetic damage of
Arthrogryposis multiplex congenita neurogenic type, genetic damage of
Arthrogryposis multiplex congenita pulmonary hypoplasia, genetic damage of
Arthrogryposis multiplex congenita whistling face, genetic damage of
Arthrogryposis ophthalmoplegia retinopathy, genetic damage of
Arthrogryposis renal dysfunction cholestasis syndrome, genetic damage of
Arthrogryposis spinal muscular atrophy, genetic damage of
Artrogriposis, genetic damage of
Ascher's syndrome, genetic damage of
Asherman's syndrome, genetic damage of
Aspartylglycosaminuria, genetic damage of

Asperger syndrome, genetic damage of
Asphyxia neonatorum, genetic damage of
Asphyxiating thoracic dystrophy, genetic damage of
Asthenia, genetic damage of
Asthenophobia, genetic damage of (possible)
Astrocytoma, genetic damage of
Asymmetric septal hypertrophy, genetic damage of
Ataxia telangiectasia, genetic damage of
Ataxia telangiectasia variant V1, genetic damage of
Ataxia, Marie's, genetic damage of
Ataxiophobia, genetic damage of (possible)
Ataxophobia, genetic damage of (possible)
Athetosis, genetic damage of
Atkin Flaitz Patil Smith syndrome, genetic damage of
ATR-X, genetic damage of
Atresia of small intestine, genetic damage of
Atrial myxoma, familial, genetic damage of
Atrial septal defects, genetic damage of
Atrioventricular septal defect, genetic damage of
Atrophoderma of Pierini and Pasini, genetic damage of
Attention Deficit Disorder with Hyperactivity, genetic damage of
Atychiphobia, genetic damage of (possible)
Atypical lipodystrophy, genetic damage of
Auditory Perceptual disorder, genetic damage of
Aughton syndrome, genetic damage of
Ausems Wittebol Post Hennekam syndrome, genetic damage of
Autism, genetic damage of
Autoimmune hemolytic anemia, genetic damage of
Autoimmune hepatitis, genetic damage of
Autoimmune peripheral neuropathy, genetic damage of
Automysophobia, genetic damage of (possible)
Autonomic dysfunction, genetic damage of
Autonomic nervous system diseases, genetic damage of

Axial mesodermal dysplasia spectrum, genetic damage of
Axial osteosclerosis, genetic damage of
Ayazi syndrome, genetic damage of

B

B-cell lymphomas, genetic damage of
Bacillophobia, genetic damage of (possible)
Bader syndrome, genetic damage of
Baelz syndrome, genetic damage of
Bagatelle Cassidy syndrome, genetic damage of
Bahemuka Brown syndrome, genetic damage of
Baker Vinters syndrome, genetic damage of
Baker-Winegard syndrome, genetic damage of
Ballard syndrome, genetic damage of
Baller-Gerold syndrome, genetic damage of
Ballinger-Wallace syndrome, genetic damage of
Ballistophobia, genetic damage of (possible)
Balo disease, genetic damage of
Balo's concentric sclerosis, genetic damage of
Bamforth syndrome, genetic damage of
BANF acoustic neurinoma, genetic damage of
Bangstad syndrome, genetic damage of
Banki syndrome, genetic damage of
Bannayan-Zonana syndrome, genetic damage of
Bantil's syndrome, genetic damage of
Baraitser Brett Piesowicz syndrome, genetic damage of
Baraitser Burn Fixen syndrome, genetic damage of
Baraitser Rodeck Garner syndrome, genetic damage of
Barber Say syndrome, genetic damage of
Bardet-Biedl syndrome, type 1, genetic damage of
Bardet-Biedl syndrome, type 2, genetic damage of
Bardet-Biedl syndrome, type 3, genetic damage of
Bardet-Biedl syndrome, type 4, genetic damage of
Bare lymphocyte syndrome, genetic damage of
Barnicoat Baraitser syndrome, genetic damage of
Barophobia, genetic damage of (possible)
Barrett esophagus, genetic damage of

Barrow Fitzsimmons syndrome, genetic damage of
Bart Pumphrey syndrome, genetic damage of
Barth syndrome, genetic damage of
Bartsocas Papa syndrome, genetic damage of
Bartter syndrome, genetic damage of
Bartter syndrome, antenatal form, genetic damge of
Bartter's disease, genetic damage of
Basal cell nevus anodontia abnormal bone mineralization, genetic damage of
Basal ganglia diseases, genetic damage of
Basan syndrome, genetic damage of
Basaran Yilmaz syndrome, genetic damage of
Basedow's coma, genetic damage of
Basilar artery migraines, genetic damage of
Basilar impression primary, genetic damage of
Bassoe syndrome, genetic damage of
Bathophobia, genetic damage of (possible)
Batrachophobia, genetic damage of (possible)
Battaglia Neri syndrome, genetic damage of
Batten disease, genetic damage of
Batten Turner muscular dystrophy, genetic damage of
Baughman syndrome, genetic damage of
Bazex-Dupre-Christol syndrome, genetic damage of
Bazopoulou Kyrkanidou syndrome, genetic damage of
Bd syndrome, genetic damage of
Beals syndrome, genetic damage of
Beardwell syndrome, genetic damage of
Bébé Collodion syndrome, genetic damage of
Becker disease, genetic damage of
Becker's muscular dystrophy, genetic damage of
Becker's nevus, genetic damage of
Beckwith-Wiedemann syndrome, genetic damage of
Beemer Ertbruggen syndrome, genetic damage of

Beemer Langer syndrome, genetic damage of
Behcet syndrome, genetic damage of
Behr syndrome, genetic damage of
Behrens Baumann Dust syndrome, genetic damage of
Bell's palsy, genetic damage of
Bellini Chiumello Rinoldi syndrome, genetic damage of
Ben Ari Shuper Mimouni syndrome, genetic damage of
Benallegue Lacete syndrome, genetic damage of
Bencze syndrome, genetic damage of
Benign astrocytoma, genetic damage of
Benign autosomal dominant myopathy, genetic damage of
Benign congenital hypotonia, genetic damage of
Benign essential blepharospasm, genetic damage of
Benign essential tremor syndrome, genetic damage of
Benign familial hematuria, genetic damage of
Benign familial infantile convulsions, genetic damage of
Benign familial infantile epilepsy, genetic damage of
Benign familial pemphigus, genetic damage of
Benign lymphoma, genetic damage of
Benign mucosal pemphigoid, genetic damage of
Benign paroxysmal positional vertigo (BPPV), genetic damage of
Bennion Patterson syndrome, genetic damage of
Bentham Driessen Hanveld syndrome, genetic damage of
Beradinelli syndrome, genetic damage of
Berardinelli-Seip congenital lipodystrophy, genetic damage of
Berdon syndrome, genetic damage of
Berger disease, genetic damage of
Berk Tabatznik syndrome, genetic damage of
Berlin Breakage syndrome, genetic damage of
Bernard-Soulier syndrome, genetic damage of
Besnier-Boeck-Schaumann disease, genetic damage of
Beta-galactosidase-1 deficiency, genetic damage of
Beta-mannosidosis, genetic damage of

Beta-sarcoglycanopathy, genetic damage of
Beta-thalassemia, genetic damage of
Beta-thalassemia major anemia, genetic damage of
Betaketothiolase deficiency, genetic damage of
Bethlem myopathy, genetic damage of
Bhaskar Jagannathan syndrome, genetic damage of
Bibliophobia, genetic damage of (possible)
Bickel Fanconi glycogenosis, genetic damage of
Bicuspid aortic valve, genetic damage of
Bidirectional tachycardia, genetic damage of
BIDS syndrome, genetic damage of
Biemond syndrome, genetic damage of
Biemond syndrome type 1, genetic damage of
Biemond syndrome type 2, genetic damage of
Biermer disease, genetic damage of
Bifid nose dominant, genetic damage of
Bilateral acoustic neurofibromatosis, genetic damage of
Bilateral renal agenesis (Potter syndrome), genetic damage of
Bilateral renal agenesis dominant type, genetic damage of
Biliary atresia (generic term), genetic damage of
Biliary atresia, extrahepatic, genetic damage of
Biliary atresia, intrahepatic, non syndromic form, genetic damage of
Biliary atresia, intrahepatic, syndromic form, genetic damage of
Biliary cirrhosis, genetic damage of
Biliary malformation renal tubular insufficiency, genetic damage of
Biliary tract cancer, genetic damage of
Billard Toutain Maheut syndrome, genetic damage of
Billet Bear syndrome, genetic damage of
Bindewald Ulmer Muller syndrome, genetic damage of
Binswanger's disease, genetic damage of
Biotinidase deficiency, genetic damage of

Bird headed dwarfism Montreal type, genetic damage of
Birdshot chorioretinopathy, genetic damage of
Bixler Christian Gorlin syndrome, genetic damage of
Bjornstad syndrome, genetic damage of
Blackfan-Diamond anemia, genetic damage of
Bladder cancer, genetic damage of
Bladder neoplasm, genetic damage of
Blaichman syndrome, genetic damage of
Blastoma, genetic damage of
Blepharo cheilo dontic syndrome, genetic damage of
Blepharo facio skeletal syndrome, genetic damage of
Blepharo naso facial syndrome Van maldergem type, genetic damage of
Blepharonasofacial malformation syndrome, genetic damage of
Blepharophimosis, genetic damage of
Blepharophimosis nasal groove growth retardation, genetic damage of
Blepharophimosis ptosis esotropia syndactyly short, genetic damage of
Blepharophimosis ptosis syndactyly mental retardation, genetic damage of
Blepharophimosis radioulnar synostosis, genetic damage of
Blepharophimosis syndrome Ohdo type, genetic damage of
Blepharophimosis, ptosis, epicanthus inversus, genetic damage of
Blepharoptosis aortic anomaly, genetic damage of
Blepharoptosis cleft palate ectrodactyly dental anomalies, genetic damage of
Blepharoptosis myopia ectopia lentis, genetic damage of
Blepharospasm, genetic damage of
Blethen Wenick Hawkins syndrome, genetic damage of
Blomstrand syndrome, genetic damage of
Blood platelet disorders, genetic damage of
Blood vessel disorder, genetic damage of

Bloom syndrome, genetic damage of
Blount disease, genetic damage of
Blue cone monochromatism, genetic damage of
Blue diaper syndrome, genetic damage of
Blue rubber bleb nevus, genetic damage of
Bod syndrome, genetic damage of
Bone development disorder, genetic damage of
Bone dysplasia Azouz type, genetic damage of
Bone dysplasia corpus callosum agenesis, genetic damage of
Bone dysplasia lethal Holmgren type, genetic damage of
Bone dysplasia Moore type, genetic damage of
Bone fragility craniosynostosis proptosis hydrocephalus, genetic damage of
Bone marrow failure, genetic damage of
Bone marrow failure neurologic abnormalities, genetic damage of
Bone neoplasms, genetic damage of
Bone tumor (generic term), genetic damage of
Bonneau-Beaumont syndrome, genetic damage of
Bonneman Meinecke Reich syndrome, genetic damage of
Bonnemann Meinecke syndrome, genetic damage of
Bonnevie Ullrich Turner syndrome, genetic damage of
Book syndrome, genetic damage of
Boomerang dysplasia, genetic damage of
Booth Haworth Dilling syndrome, genetic damage of
BOR syndrome, genetic damage of
Borjeson syndrome, genetic damage of
Borjeson-Forssman-Lehmann syndrome, genetic damage of
Bork Stender Schmidt syndrome, genetic damage of
Borrone Di Rocco Crovato syndrome, genetic damage of
Boscherini Galasso Manca Bitti syndrome, genetic damage of
Bosma Henkin Christiansen syndrome, genetic damage of
Boucher Neuhauser syndrome, genetic damage of
Boudhina Yedes Khiari syndrome, genetic damage of

Bourneville syndrome, genetic damage of
Bourneville syndrome, type 1, genetic damage of
Bourneville syndrome, type 2, genetic damage of
Bowen syndrome, genetic damage of
Bowen's disease, genetic damage of
Bowen-Conradi syndrome, genetic damage of
Bowing congenital short bones, genetic damage of
Bowing of long bones congenital, genetic damage of
Boylan Dew Greco syndrome, genetic damage of
BPPV (benign paroxysmal positional vertigo), genetic damage of
Brachioskeletogenital syndrome, genetic damage of
Brachman-de Lange syndrome, genetic damage of
Brachycephalofrontonasal dysplasia, genetic damage of
Brachycephaly deafness cataract mental retardation, genetic damage of
Brachydactylie types b and e combined, genetic damage of
Brachydactylous dwarfism Mseleni type, genetic damage of
Brachydactyly absence of distal phalanges, genetic damage of
Brachydactyly anonychia, genetic damage of
Brachydactyly clinodactyly, genetic damage of
Brachydactyly dwarfism mental retardation, genetic damage of
Brachydactyly elbow wrist dysplasia, genetic damage of
Brachydactyly hypertension, genetic damage of
Brachydactyly long thumb type, genetic damage of
Brachydactyly mesomelia mental retardation heart defects, genetic damage of
Brachydactyly Mohr Wriedt type, genetic damage of
Brachydactyly nystagmus cerebellar ataxia, genetic damage of
Brachydactyly preaxial hallux varus, genetic damage of
Brachydactyly scoliosis carpal fusion, genetic damage of
Brachydactyly small stature face anomalies, genetic damage of
Brachydactyly Smorgasbord type, genetic damage of
Brachydactyly Temtamy type, genetic damage of

Brachydactyly tibial hypoplasia, genetic damage of
Brachydactyly type a1, genetic damage of
Brachydactyly type a2, genetic damage of
Brachydactyly type a3, genetic damage of
Brachydactyly type a4, genetic damage of
Brachydactyly type a5 nail dysplasia, genetic damage of
Brachydactyly type a6, genetic damage of
Brachydactyly type a7, genetic damage of
Brachydactyly type b, genetic damage of
Brachydactyly type c, genetic damage of
Brachydactyly type e, genetic damage of
Brachymesomelia renal syndrome, genetic damage of
Brachymesophalangy 2 and 5, genetic damage of
Brachymesophalangy mesomelic short limbs osseous anomalies, genetic damage of
Brachymesophalangy type 2, genetic damage of
Brachymetapody anodontia hypotrichosis albinoidism, genetic damage of
Brachymorphism onychodysplasia dysphalangism syndrome, genetic damage of
Brachyolmia, genetic damage of
Brachyolmia recessive Hobaek type, genetic damage of
Brachytelephalangy characteristic facies Kallmann, genetic damage of
Braddock Carey syndrome, genetic damage of
Braddock Jones Superneau syndrome, genetic damage of
Bradykinesia, genetic damage of
Brain cavernous angioma, genetic damage of
Brain neoplasms, genetic damage of
Brain stem neoplasms, genetic damage of
Branched chain ketoaciduria, genetic damage of
Branchial arch defects, genetic damage of
Branchial arch syndrome X linked, genetic damage of

Branchial dysplasia mental retardation inguinal hernia, genetic damage of
Branchio oculo facial syndrome Hing type, genetic damage of
Branchio-oculo-facial syndrome, genetic damage of
Branchiootorenal syndrome, genetic damage of
Breast and ovarian cancer, genetic damage of
Breast cancer, familial, genetic damage of
Brittle bone disease, genetic damage of
Brittle bone syndrome lethal type, genetic damage of
Brittle cornea syndrome, genetic damage of
Broad beta disease, genetic damage of
Broad-betalipoproteinemia, genetic damage of
Bromidrosiphobia, genetic damage of (possible)
Bronchiectasis oligospermia, genetic damage of
Bronchiolitis obliterans with obstructive pulmonary disease, genetic damage of
Bronchiolotis obliterans organizing pneumonia (BOOP), genetic damage of
Bronchogenic cyst, genetic damage of
Bronchopulmonary amyloidosis, genetic damage of
Bronchopulmonary dysplasia, genetic damage of
Brown syndrome, genetic damage of
Brown-Sequard syndrome, genetic damage of
Bruce Winship syndrome, genetic damage of
Brucellosis, genetic damage of
Bruck syndrome, genetic damage of
Brugada syndrome, genetic damage of
Brunner Winter syndrome, genetic damage of
Brunoni syndrome, genetic damage of
Bruton type agammaglobulinemia, genetic damage of
Bruyn Scheltens syndrome, genetic damage of
Budd-Chiari syndrome, genetic damage of
Buerger's disease, genetic damage of

Bufonophobia, genetic damage of (possible)
Bulbospinal amyotrophy, X linked, genetic damage of
Bulimia nervosa, genetic damage of
Bull Nixon syndrome, genetic damage of
Bullous dystrophy macular type, genetic damage of
Bullous ichtyosiform erythroderma congenita, genetic damage of
Bullous pemphigoid, genetic damage of
Buntinx Lormans Martin syndrome, genetic damage of
Burkitt's lymphoma, genetic damage of
Burn Goodship syndrome, genetic damage of
Burnett Schwartz Berberian syndrome, genetic damage of
Burning mouth syndrome, genetic damage of
Buschke Fischer Brauer syndrome, genetic damage of
Buschke Ollendorff syndrome, genetic damage of
Buschke's scleredema, genetic damage of
Bustos Simosa Pinto Cisternas syndrome, genetic damage of
Buttiens Fryns syndrome, genetic damage of
Butyrylcholinesterase deficiency, genetic damage of

C

C syndrome, genetic damage of
C1 esterase deficiency (type 2 with ascites), genetic damage of
Cacchi Ricci disease, genetic damage of
CACH syndrome, genetic damage of
Cacophobia, genetic damage of (possible)
CADASIL, genetic damage of
Cafe au lait spots syndrome, genetic damage of
Caffey disease, genetic damage of
Cahmr syndrome, genetic damage of
Calcinosis-Raynaud   phenomenon-sclerodactyly-telangiectasia, genetic damage of
Calciphylaxis, genetic damage of
Calculi, genetic damage of (possible)
Calderon Gonzalez Cantu syndrome, genetic damage of
Calloso genital dysplasia, genetic damage of
Callus disease, genetic damage of
Calpainopathy, genetic damage of
Calvarial hyperostosis, genetic damage of
Camera Lituania Cohen syndrome, genetic damage of
Camfak syndrome, genetic damage of
Campomelia Cumming type, genetic damage of
Camptobrachydactyly, genetic damage of
Camptocormism, genetic damage of
Camptodactyly fibrous tissue hyperplasia skeletal dysplasia, genetic damage of
Camptodactyly joint contractures facial skeletal dysplasia, genetic damage of
Camptodactyly overgrowth unusual facies, genetic damage of
Camptodactyly syndrome Guadalajara type 1, genetic damage of
Camptodactyly syndrome Guadalajara type 2, genetic damage of
Camptodactyly taurinuria, genetic damage of
Camptodactyly vertebral fusion, genetic damage of

Camptomelic syndrome, genetic damage of
Camurati Engelmann disease, genetic damage of
Canavan leukodystrophy, genetic damage of
Cancrum oris, genetic damage of
Cantalamessa Baldini Ambrosi syndrome, genetic damage of
Cantu Sanchez Corona Fragoso syndrome, genetic damage of
Cantu Sanchez Corona Garcia syndrome, genetic damage of
Cantu Sanchez Corona Hernandes syndrome, genetic damage of
Capillary leak syndrome with monoclonal gammopathy, genetic damage of
Capillary venous leptomeningeal angiomatosis, genetic damage of
Capos syndrome, genetic damage of
Caratolo Cilio Pessagno syndrome, genetic damage of
Carbamoyl phosphate synthetase deficiency, genetic damage of
Carbamoyl phosphate synthetase deficiency (genetic form), genetic damage of
Carbamoyl-phosphate synthase I deficiency disease (ornithine carbamoyl phosphate deficiency), genetic damage of
Carbamyl phosphate synthetase deficiency, genetic damage of
Carbohydrate deficient glycoprotein syndrome, genetic damage of
Carbon baby syndrome, genetic damage of
Carbonic anhydrase II deficiency, genetic damage of
Carboxylase deficiency, multiple, genetic damage of
Carcinoid syndrome, genetic damage of
Carcinoma, genetic damage of
Carcinoma of the vocal tract, genetic damage of
Carcinoma, squamous cell, genetic damage of
Carcinophobia, genetic damage of (possible)
Cardiac and laterality defects, genetic damage of
Cardiac conduction defect, familial, genetic damage of
Cardiac diverticulum, genetic damage of
Cardiac hydatid cysts with intracavitary expansion, genetic damage of

Cardiac malformation, genetic damage of
Cardiac valvular dysplasia, X linked, genetic damage of
Cardio-facio-cutaneous syndrome, genetic damage of
Cardioauditory syndrome, genetic damage of
Cardiofacial syndrome short limbs, genetic damage of
Cardiogenital syndrome, genetic damage of
Cardiomelic syndrome Stratton Koehler type, genetic damage of
Cardiomyopathic lentiginosis, genetic damage of (possible)
Cardiomyopathy cataract hip spine disease, genetic damage of
Cardiomyopathy diabetes deafness, genetic damage of
Cardiomyopathy dilated with conduction defect type 1, genetic damage of
Cardiomyopathy dilated with conduction defect type 2, genetic damage of
Cardiomyopathy due to anthracyclines, genetic damage of
Cardiomyopathy hearing loss type t RNA lysine gene mutation, genetic damage of
Cardiomyopathy hypogonadism metabolic anomalies, genetic damage of
Cardiomyopathy spherocytosis, genetic damage of
Cardiomyopathy, familial dilated, genetic damage of
Cardiomyopathy, familial hypertrophic, genetic damage of
Cardiomyopathy, X linked, fatal infantile, genetic damage of
Cardiophobia, genetic damage of (possible)
Cardioskeletal myopathy-neutropenia, genetic damage of
Cardiospasm, genetic damage of
Carey Fineman Ziter syndrome, genetic damage of
Carnevale Canun Mendoza syndrome, genetic damage of
Carnevale Hernandez Castillo syndrome, genetic damage of
Carnevale Krajewska Fischetto syndrome, genetic damage of
Carney syndrome, genetic damage of
Carnitine palmitoyl transferase 1 deficiency, genetic damage of
Carnitine palmitoyl transferase 2 deficiency, genetic damage of

Carnitine palmityl transferase deficiency, genetic damage of
Carnitine transporter deficiency, genetic damage of
Carnitine-acylcarnitine translocase deficiency, genetic damage of
Carnophobia, genetic damage of (possible)
Carnosinase deficiency, genetic damage of
Carnosinemia, genetic damage of
Caroli disease, genetic damage of
Carpal deformity migrognathia microstomia, genetic damage of
Carpal tunnel syndrome, genetic damage of
Carpenter Hunter type, genetic damage of
Carpenter syndrome, genetic damage of
Carpo tarsal osteochondromatosis, genetic damage of
Carpo tarsal osteolysis recessive, genetic damage of
Carrington syndrome, genetic damage of
Cartilage hair hypoplasia like syndrome, genetic damage of
Cartilaginous neoplasms, genetic damage of
Cartwright Nelson Fryns syndrome, genetic damage of
Cassia Stocco Dos Santos syndrome, genetic damage of
Castleman's disease, genetic damage of
Castro Gago Pombo Novo syndrome, genetic damage of
Cat cry syndrome, genetic damage of
Cat eye syndrome, genetic damage of
Cat Rodrigues syndrome, genetic damage of
Cat scratch disease, genetic damage of
Catagelophobia, genetic damage of (possible)
Catapedaphobia, genetic damage of (possible)
Cataract aberrant oral frenula growth retardation, genetic damage of
Cataract alopecia sclerodactyly, genetic damage of
Cataract anterior polar dominant, genetic damage of
Cataract ataxia deafness, genetic damage of
Cataract cardiomyopathy, genetic damage of
Cataract congenital autosomal dominant, genetic damage of

Cataract congenital dominant non nuclear, genetic damage of
Cataract congenital ichthyosis, genetic damage of
Cataract congenital Volkmann type, genetic damage of
Cataract congenital with microphthalmia, genetic damage of
Cataract dental syndrome, genetic damage of
Cataract Hutterite type, genetic damage of
Cataract hypertrichosis mental retardation, genetic damage of
Cataract mental retardation hypogonadism, genetic damage of
Cataract microcornea syndrome, genetic damage of
Cataract microcornea X linked, genetic damage of
Cataract microphthalmia septal defect, genetic damage of
Cataract skeletal anomalies, genetic damage of
Cataract, total congenital, genetic damage of
Cataract-glaucoma, genetic damage of
CATCH 22 syndrome, genetic damage of
Catecholamine hypertension, genetic damage of
Catel Manzke syndrome, genetic damage of
Caudal appendage deafness, genetic damage of
Caudal duplication, genetic damage of
Caudal regression syndrome, genetic damage of
Causalgia, genetic damage of
Cavernous hemangioma, genetic damage of
Cavernous lymphangioma, genetic damage of
Cayler syndrome, genetic damage of
CBPS or Perisylvian syndrome, genetic damage of
CCA syndrome, genetic damage of
Ccge syndrome, genetic damage of
CDG syndrome, genetic damage of
CDG syndrome type 1A, genetic damage of
CDG syndrome type 1B, genetic damage of
CDG syndrome type 1C, genetic damage of
CDG syndrome type 2, genetic damage of
CDG syndrome type 3, genetic damage of

CDG syndrome type 4, genetic damage of
CDK4 linked melanoma, genetic damage of
Cecato De Lima Pinheiro syndrome, genetic damage of
Celiac disease epilepsy occipital calcifications, genetic damage of
Celiac sprue, genetic damage of
Cenani Lenz syndactylism, genetic damage of
Cennamo Gangemi syndrome, genetic damage of
Central core disease, genetic damage of
Central diabetes insipidus, genetic damage of
Central serous chorioretinopathy, genetic damage of
Central type neurofibromatosis, genetic damage of
Centromeric instability immunodeficiency syndrome, genetic damage of
Centronuclear myopathy, congenital, genetic damage of
Centrotemporal epilepsy, genetic damage of
Cephalopolysyndactyly, genetic damage of
Ceramidase deficiency, genetic damage of
Ceramide trihexosidosis, genetic damage of
Ceraunophobia, genetic damage of (possible)
Cerebellar agenesis, genetic damage of
Cerebellar ataxia, genetic damage of
Cerebellar ataxia areflexia pes cavus optic atrophy, genetic damage of
Cerebellar ataxia ectodermal dysplasia, genetic damage of
Cerebellar ataxia infantile with progressive external opht halmoplegia, genetic damage of
Cerebellar ataxia, dominant pure, genetic damage of
Cerebellar degeneration, genetic damage of
Cerebellar degeneration, subacute, genetic damage of
Cerebellar hypoplasia, genetic damage of
Cerebellar hypoplasia endosteal sclerosis, genetic damage of
Cerebellar hypoplasia tapetoretinal degeneration, genetic damage of

Cerebellar parenchymal degeneration, genetic damage of
Cerebellar vermis agenesis, genetic damage of
Cerebelloolivary atrophy, genetic damage of
Cerebelloparenchymal disorder 3, genetic damage of
Cerebellum agenesis hydrocephaly, genetic damage of
Cerebral amyloid angiopathy, genetic damage of
Cerebral aneurysm, genetic damage of
Cerebral calcification cerebellar hypoplasia, genetic damage of
Cerebral calcifications opalescent teeth phosphaturia, genetic damage of
Cerebral cavernous malformation, genetic damage of
Cerebral cavernous malformations, genetic damage of
Cerebral gigantism, genetic damage of
Cerebral gigantism jaw cysts, genetic damage of
Cerebral malformations hypertrichosis claw hands, genetic damage of
Cerebral palsy, genetic damage of
Cerebral ventricle neoplasm, genetic damage of
Cerebro facio articular syndrome, genetic damage of
Cerebro facio thoracic dysplasia, genetic damage of
Cerebro oculo dento auriculo skeletal syndrome, genetic damage of
Cerebro oculo facio skeletal syndrome, genetic damage of
Cerebro oculo genital syndrome, genetic damage of
Cerebro oculo skeleto renal syndrome, genetic damage of
Cerebro reno digital syndrome, genetic damage of
Cerebro-costo-mandibular syndrome, genetic damage of
Cerebro-oculo-facio-skeletal syndrome, genetic damage of
Cerebroarthrodigital syndrome, genetic damage of
Cerebroretinal vasculopathy, genetic damage of
Cerebrotendinous xanthomatosis (CTX), genetic damage of
Ceroid lipofuscinose, neuronal, genetic damage of
Ceroid lipofuscinose, neuronal 1, infantile, genetic damage of

Ceroid lipofuscinose, neuronal 2, late infantile, genetic damage of
Ceroid lipofuscinose, neuronal 3, juvenile, genetic damage of
Ceroid lipofuscinose, neuronal 4, adult type, genetic damage of
Ceroid lipofuscinose, neuronal 5, late infantile, genetic damage of
Ceroid lipofuscinose, neuronal 6, late infantile, genetic damage of
Cervical hypertrichosis neuropathy, genetic damage of
Cervical hypertrichosis peripheral neuropathy, genetic damage of
Cervical ribs sprengel anomaly polydactyly, genetic damage of
Cervical vertebral fusion, genetic damage of
Cervicooculoacoustic syndrome, genetic damage of
CFC syndrome, genetic damage of
Chagas disease, genetic damage of
Chalazion, genetic damage of
Chanarin disease, genetic damage of
Chanarin Dorfman syndrome ichthyosis, genetic damage of
Chandler's syndrome, genetic damage of
Chands syndrome, genetic damage of
Chang Davidson Carlson syndrome, genetic damage of
Chaotic atrial tachycardia, genetic damage of
Char syndrome, genetic damage of
Charcot disease, genetic damage of
Charcot Marie tooth disease deafness dominant type, genetic damage of
Charcot Marie tooth disease deafness mental retardation, genetic damage of
Charcot Marie tooth disease deafness recessive type, genetic damage of
Charcot Marie tooth type 1 aplasia cutis congenita, genetic damage of
Charcot-Marie-tooth disease, genetic damage of
Charcot-Marie-tooth disease, X linked type 2, recessive, genetic damage of
Charcot-Marie-tooth disease, X linked type 3, recessive, genetic

damage of

Charcot-Marie-tooth disease type 1A, genetic damage of

Charcot-Marie-tooth disease type 1B, genetic damage of

Charcot-Marie-tooth disease type 1C, genetic damage of

Charcot-Marie-tooth disease type 2A, genetic damage of

Charcot-Marie-tooth disease type 2B, genetic damage of

Charcot-Marie-tooth disease type 2C, genetic damage of

Charcot-Marie-tooth disease type 2D, genetic damage of

Charcot-Marie-tooth disease type 4A, genetic damage of

Charcot-Marie-tooth disease, intermediate form, genetic damage of

Charcot-Marie-tooth disease, neuronal, type A, genetic damage of

Charcot-Marie-tooth disease, neuronal, type B, genetic damage of

Charcot-Marie-tooth disease, neuronal, type C, genetic damage of

Charcot-Marie-tooth disease, neuronal, type D, genetic damage of

Charcot-Marie-tooth peroneal muscular atrophy, X linked, genetic damage of

CHARGE association, genetic damage of

Charlie M syndrome, genetic damage of

Chavany-Brunhes syndrome, genetic damage of

Chediak-Higashi syndrome, genetic damage of

Chediak-Higashi-like syndrome, genetic damage of

Chemke Oliver Mallek syndrome, genetic damage of

Chemodectoma, genetic damage of

Chemophobia, genetic damage of (possible)

Chen Kung Ho Kaufman Mcalister syndrome, genetic damage of

Cherubism, genetic damage of

Cherubism gingival fibromatosis mental retardation, genetic damage of

Cherubism optic atrophy short stature, genetic damage of

Chiari type 1 malformation, genetic damage of

Chiari-Frommel syndrome, genetic damage of

CHILD syndrome ichthyosis, genetic damage of

Childhood ataxia with diffuse central nervous system, genetic damage of
Childhood disintegrative disorder, genetic damage of
Childhood pustular psoriasis, genetic damage of
Chionophobia, genetic damage of (possible)
Chiraptophobia, genetic damage of (possible)
Chirophobia, genetic damage of (possible)
Chitayat Haj Chahine syndrome, genetic damage of
Chitayat Meunier Hodgkinson syndrome, genetic damage of
Chitayat Moore Del Bigio syndrome, genetic damage of
Chitty Hall Baraitser syndrome, genetic damage of
Chitty Hall Webb syndrome, genetic damage of
Choanal atresia deafness cardiac defects dysmorphia, genetic damage of
Cholangiocarcinoma, genetic damage of
Cholangitis, primary sclerosing, genetic damage of
Cholecystitis, genetic damage of
Choledochal cyst hand malformation, genetic damage of
Cholemia, familial, genetic damage of
Cholerophobia, genetic damage of (possible)
Cholestasis, genetic damage of
Cholestasis pigmentary retinopathy cleft palate, genetic damage of
Cholestasis, progressive familial intrahepatic, genetic damage of
Cholestasis, progressive familial intrahepatic 1, genetic damage of
Cholestasis, progressive familial intrahepatic 2, genetic damage of
Cholestasis, progressive familial intrahepatic 3, genetic damage of
Cholestatic jaundice renal tubular insufficiency, genetic damage of
Cholesterol ester storage disease, genetic damage of
Cholesterol esterification disorder, genetic damage of
Chondroblastoma (benign), genetic damage of
Chondrocalcinosis, genetic damage of
Chondrocalcinosis familial articular, genetic damage of
Chondrodysplasia lethal recessive, genetic damage of

Chondrodysplasia pseudohermaphrodism syndrome, genetic damage of
Chondrodysplasia punctata, genetic damage of
Chondrodysplasia punctata with steroid sulfatase deficiency, genetic damage of
Chondrodysplasia punctata, brachytelephalangic, genetic damage of
Chondrodysplasia punctata, rhizomelic form, genetic damage of
Chondrodysplasia punctata, Sheffield type, genetic damage of
Chondrodysplasia situs inversus imperforate anus polydactyly, genetic damage of
Chondrodysplasia, Grebe type, genetic damage of
Chondrodystrophy, genetic damage of
Chondroectodermal dysplasia, genetic damage of
Chondroma (benign), genetic damage of
Chondromalacia, genetic damage of
Chondromatosis (benign), genetic damage of
Chondrosarcoma (malignant), genetic damage of
Chondrysplasia punctata, humero-metacarpal type, genetic damage of
Chordoma, genetic damage of
Chorea, genetic damage of
Chorea acanthocytosis, genetic damage of
Chorea familial benign, genetic damage of
Chorea minor, genetic damage of
Choreoacanthocytosis amyotrophic, genetic damage of
Choreoathetosis familial paroxysmal, genetic damage of
Choreoathetosis hyperuricemia, genetic damage of
Choriocarcinoma, genetic damage of
Chorioretinopathy dominant form microcephaly, genetic damage of
Choroid plexus cyst, genetic damage of
Choroid plexus neoplasms, genetic damage of

Choroidal atrophy alopecia, genetic damage of
Choroideremia, genetic damage of
Choroideremia hypopituitarism, genetic damage of
Choroiditis, genetic damage of
Choroiditis, serpiginous, genetic damage of
Choroido cerebral calcification syndrome infantile, genetic damage of
Chorophobia, genetic damage of (possible)
Christian Demyer Franken syndrome, genetic damage of
Christian Johnson Angenieta syndrome, genetic damage of
Christian syndrome, genetic damage of
Christianson Fourie syndrome, genetic damage of
Christmas disease, genetic damage of
Chrometophobia, genetic damage of (possible)
Chromophobe renal carcinoma, genetic damage of
Chromophobia, genetic damage of (possible)
Chromosomal triplication, genetic damage of
Chromosome 1 ring, genetic damage of
Chromosome 10 ring, genetic damage of
Chromosome 10, distal trisomy 10q, genetic damage of
Chromosome 10, monosomy 10p, genetic damage of
Chromosome 11, partial monosomy 11q, genetic damage of
Chromosome 11, partial trisomy 11q, genetic damage of
Chromosome 11–14 translocation, genetic damage of
Chromosome 11q syndrome, genetic damage of
Chromosome 12 ring, genetic damage of
Chromosome 12p deletion, genetic damage of
Chromosome 13 ring, genetic damage of
Chromosome 13, partial monosomy 13q, genetic damage of
Chromosome 13q syndrome, genetic damage of
Chromosome 13q-mosaicism, genetic damage of
Chromosome 14 ring, genetic damage of
Chromosome 14 trisomy, genetic damage of

Chromosome 14, trisomy mosaic, genetic damage of
Chromosome 15 ring, genetic damage of
Chromosome 15, distal trisomy 15q, genetic damage of
Chromosome 17deletion, genetic damage of
Chromosome 18 long arm deletion syndrome, genetic damage of
Chromosome 18, monosomy 18p, genetic damage of
Chromosome 18, ring, genetic damage of
Chromosome 18, tetrasomy 18p, genetic damage of
Chromosome 18p-syndrome, genetic damage of
Chromosome 18q-syndrome, genetic damage of
Chromosome 19 ring, genetic damage of
Chromosome 1p36 depletion syndrome, genetic damage of
Chromosome 20 ring, genetic damage of
Chromosome 21 ring, genetic damage of
Chromosome 22 ring, genetic damage of
Chromosome 22 trisomy, genetic damage of
Chromosome 22, trisomy mosaic, genetic damage of
Chromosome 3 deletion of distal, genetic damage of
Chromosome 3 duplication syndrome, genetic damage of
Chromosome 3, monosomy 3p, genetic damage of
Chromosome 3, monosomy 3p2, genetic damage of
Chromosome 3, Trisomy 3q2, genetic damage of
Chromosome 4 ring, genetic damage of
Chromosome 4 short arm deletion, genetic damage of
Chromosome 4 Trisomy, genetic damage of
Chromosome 4, monosomy 4q, genetic damage of
Chromosome 4, monosomy distal 4q, genetic damage of
Chromosome 4, Partial Trisomy Distal 4q, genetic damage of
Chromosome 4, Trisomy 4p, genetic damage of
Chromosome 4q-syndrome, genetic damage of
Chromosome 5 trisomy 5p, genetic damage of
Chromosome 5p-syndrome, genetic damage of
Chromosome 6 ring, genetic damage of

Chromosome 6, partial trisomy 6q, genetic damage of
Chromosome 7 ring, genetic damage of
Chromosome 7, monosomy 7p2, genetic damage of
Chromosome 8 deletion, genetic damage of
Chromosome 8 ring, genetic damage of
Chromosome 8, monosomy 8p2, genetic damage of
Chromosome 9 ring, genetic damage of
Chromosome 9, partial monosomy 9p, genetic damage of
Chromosome 9, tetrasomy 9p, genetic damage of
Chromosome 9, trisomy 9p (multiple variants), genetic damage of
Chromosome 9, trisomy mosaic, genetic damage of
Chromosome disorders, genetic damage of
Chromosome triploidy syndrome, genetic damage of
Chronic demyelinizing neuropathy with IgM monoclonal, genetic damage of
Chronic erosive gastritis, genetic damage of
Chronic fatigue immune dysfunction syndrome, genetic damage of
Chronic granulomatous disease, genetic damage of
Chronic hiccup, genetic damage of (possible)
Chronic inflammatory demyelinating polyneuropathy, genetic damage of
Chronic lymphocytic leukemia, genetic damage of
Chronic myelogenous leukemia, genetic damage of
Chronic myelomonocytic leukemia, genetic damage of
Chronic necrotizing vasculitis, genetic damage of
Chronic neutropenia, genetic damage of
Chronic polyradiculoneuritis, genetic damage of
Chronic recurrent multifocal osteomyelitis, genetic damage of
Chronic renal failure, genetic damage of
Chronic spasmodic dysphonia, genetic damage of
Chronic, infantile, neurological, cutaneous, articular syndrome, genetic damage of

Chronomentrophobia, genetic damage of (possible)
Chudley Lowry Hoar syndrome, genetic damage of
Chudley Rozdilsky syndrome, genetic damage of
Chudley-Mccullough syndrome, genetic damage of
Churg-Strauss syndrome, genetic damage of
Chylous ascites, genetic damage of
Cicatricial pemphigoid, genetic damage of
Ciliary discoordination, due to random ciliary orientation, genetic
damage of
Ciliary dyskinesia, due to transposition of ciliary microtubules,
genetic damage of
Ciliary dyskinesia-bronchiectasis, genetic damage of
Cilliers Beighton syndrome, genetic damage of
Cinca syndrome, genetic damage of
Circumscribed cutaneous aplasia of the vertex, genetic damage of
Circumscribed disseminated keratosis Jadassohn Lew type, genetic
damage of
Citrullinemia, genetic damage of
Clarkson disease, genetic damage of
Clayton Smith Donnai syndrome, genetic damage of
Cleft hand absent tibia, genetic damage of
Cleft lip, genetic damage of
Cleft lip and palate malrotation cardiopathy, genetic damage of
Cleft lip and/or palate with mucous cysts of lower lip, genetic
damage of
Cleft lip palate abnormal thumbs microcephaly, genetic damage of
Cleft lip palate deafness sacral lipoma, genetic damage of
Cleft lip palate dysmorphism Kumar type, genetic damage of
Cleft lip palate ectrodactyly, genetic damage of
Cleft lip palate facial eye heart intestinal anomalies, genetic dam-
age of
Cleft lip palate incisor and finger anomalies, genetic damage of
Cleft lip palate lip pits limb deficiency, genetic damage of

Cleft lip palate mental retardation corneal opacity, genetic damage of

Cleft lip palate oligodontia syndactyly pili torti, genetic damage of

Cleft lip palate pituitary deficiency, genetic damage of

Cleft lip palate-tetraphocomelia, genetic damage of

Cleft lip with or without cleft palate, genetic damage of

Cleft lower lip cleft lateral canthi chorioretinal, genetic damage of

Cleft palate, genetic damage of

Cleft palate cardiac defect ectrodactyly, genetic damage of

Cleft palate colobomata radial synostosis deafness, genetic damage of

Cleft palate heart disease polydactyly absent tibia, genetic damage of

Cleft palate lateral synechia syndrome, genetic damage of

Cleft palate short stature vertebral anomalies, genetic damage of

Cleft palate stapes fixation oligodontia, genetic damage of

Cleft palate X linked, genetic damage of

Cleft tongue syndrome, genetic damage of

Cleft upper lip median cutaneous polyps, genetic damage of

Clefting ectropion conical teeth, genetic damage of

Cleido rhizomelic syndrome, genetic damage of

Cleidocranial dysplasia, genetic damage of

Cleidocranial dysplasia micrognathia absent thumbs, genetic damage of

Cleisiophobia, genetic damage of (possible)

Climacophobia, genetic damage of (possible)

Clinophobia, genetic damage of (possible)

Cloacal exstrophy, genetic damage of

Clouston syndrome, genetic damage of

Cloverleaf skull bone dysplasia, genetic damage of

Cloverleaf skull micromelia thoracic dysplasia, genetic damage of

Cloverleaf skull syndrome, genetic damage of

Cluster headache, genetic damage of

Coach syndrome, genetic damage of
Coarctation of aorta dominant, genetic damage of
Coarse face hypotonia constipation, genetic damage of
Coats disease, genetic damage of
Cochin Jewish disorder, genetic damage of
Cockayne syndrome type 1, genetic damage of
Cockayne syndrome type 2, genetic damage of
Cockayne syndrome type 3, genetic damage of
Cockayne's syndrome, genetic damage of
Codas syndrome, genetic damage of
Coenzyme Q cytochrome c reductase, deficiency of, genetic damage of
Coffin-Lowry syndrome, genetic damage of
Coffin-Siris syndrome, genetic damage of
Cofs syndrome, genetic damage of
Cogan's syndrome, genetic damage of
Cogan-Reese syndrome, genetic damage of
Cohen Hayden syndrome, genetic damage of
Cohen Lockood Wyborney syndrome, genetic damage of
Cohen syndrome, genetic damage of
Coimetrophobia, genetic damage of (possible)
Colangite esclerosante por paracoccidiodomicose, genetic damage of
Colavita Kozlowski syndrome, genetic damage of
Cold agglutination syndrome, genetic damage of
Cold agglutinin disease, genetic damage of
Cold antibody hemolytic anemia, genetic damage of
Cold contact urticaria, genetic damage of
Cold urticaria, genetic damage of
Cole carpenter syndrome, genetic damage of
Coleman Randall syndrome, genetic damage of
Colitis, genetic damage of (possible)
Collagen disorder, genetic damage of

Collagenous colitis, genetic damage of
Collins Pope syndrome, genetic damage of
Collins Sakati syndrome, genetic damage of
Coloboma chorioretinal cerebellar vermis aplasia, genetic damage of
Coloboma hair abnormality, genetic damage of
Coloboma of choroid and retina, genetic damage of
Coloboma of eye lens, genetic damage of
Coloboma of iris, genetic damage of
Coloboma of lens ala nasi, genetic damage of
Coloboma of macula, genetic damage of
Coloboma of macula type b brachydactyly, genetic damage of
Coloboma of optic papilla, genetic damage of
Coloboma porencephaly hydronephrosis, genetic damage of
Coloboma uveal with cleft lip palate and mental retardation, genetic damage of
Coloboma, ocular, genetic damage of
Colobomata unilobar lung heart defect, genetic damage of
Colobomatous microphthalmia, genetic damage of
Colobomatous microphthalmia heart disease hearing, genetic damage of
Colon cancer, familial nonpolyposis, genetic damage of
Colonic atresia, genetic damage of
Colver Steer Godman syndrome, genetic damage of
Combarros Calleja Leno syndrome, genetic damage of
Combined hyperlipidemia, familial, genetic damage of
Common mesentery, genetic damage of
Common variable immunodeficiency, genetic damage of
Compartment syndrome, genetic damage of
Complement component 2 deficiency, genetic damage of
Complement component receptor 1, genetic damage of
Complete atrioventricular canal, genetic damage of
Complex 1 mitochondrial respiratory chain deficiency, genetic

damage of

Complex 2 mitochondrial respiratory chain deficiency, genetic damage of

Complex 3 mitochondrial respiratory chain deficiency, genetic damage of

Complex 4 mitochondrial respiratory chain deficiency, genetic damage of

Complex 5 mitochondrial respiratory chain deficiency, genetic damage of

Conductive deafness malformed external ear, genetic damage of

Conductive hearing loss, genetic damage of, genetic damage of

Cone dystrophy, genetic damage of

Cone rod dystrophy, genetic damage of

Cone rod dystrophy amelogenesis imperfecta, genetic damage of

Congenital absence of the uterus and vagina, genetic damage of

Congenital adrenal hyperplasia type 1, genetic damage of

Congenital adrenal hyperplasia type 2, genetic damage of

Congenital adrenal hyperplasia type 3, genetic damage of

Congenital adrenal hyperplasia type 4, genetic damage of

Congenital adrenal hyperplasia type 5, genetic damage of

Congenital afibrinogenemia, genetic damage of

Congenital alopecia X linked, genetic damage of

Congenital amputation, genetic damage of

Congenital aneurysms of the great vessels, genetic damage of

Congenital antithrombin III deficiency, genetic damage of

Congenital aplastic anemia, genetic damage of

Congenital arteriovenous shunt, genetic damage of

Congenital articular rigidity, genetic damage of

Congenital benign spinal muscular atrophy dominant, genetic damage of

Congenital brain disorder, genetic damage of

Congenital bronchobiliary fistula, genetic damage of

Congenital cardiovascular disorder, genetic damage of

Congenital cardiovascular malformations, genetic damage of
Congenital cardiovascular shunt, genetic damage of
Congenital constricting band, genetic damage of
Congenital contractual arachnodactyly, genetic damage of
Congenital contractures, genetic damage of
Congenital craniosynostosis maternal hyperthyroiditis, genetic damage of
Congenital cystic adenomatoid malformation, genetic damage of
Congenital cystic eye multiple ocular and intracranial anomalies, genetic damage of
Congenital deafness, genetic damage of
Congenital diaphragmatic hernia, genetic damage of
Congenital erythropoietic porphyria, genetic damage of
Congenital facial diplegia, genetic damage of
Congenital fiber type disproportion, genetic damage of
Congenital gastrointestinal disorder, genetic damage of
Congenital generalized fibromatosis, genetic damage of
Congenital giant megaureter, genetic damage of
Congenital heart block, genetic damage of
Congenital heart disease ptosis hypodontia craniostosis, genetic damage of
Congenital heart disease radio ulnar synostosis mental retardation, genetic damage of
Congenital heart disorders, genetic damage of
Congenital heart septum defect, genetic damage of
Congenital hemidysplasia with ichtyosiform erythroderma and limbs defects, genetic damage of
Congenital hemolytic anemia, genetic damage of
Congenital hypomyelination neuropathy, genetic damage of
Congenital hypothyroidism, genetic damage of
Congenital hypotrichosis milia, genetic damage of
Congenital ichthyosis, genetic damage of
Congenital ichthyosis microcephalus quadriplegia, genetic damage

of
Congenital ichtyosiform erythroderma, genetic damage of
Congenital kidney disorder, genetic damage of
Congenital lobar emphysema, genetic damage of
Congenital megacolon, genetic damage of
Congenital megalo-ureter, genetic damage of
Congenital mesoblastic nephroma, genetic damage of
Congenital microvillous atrophy, genetic damage of
Congenital mitral malformation, genetic damage of
Congenital mitral stenosis, genetic damage of
Congenital muscular dystrophy syringomyelia, genetic damage of
Congenital myopathy, genetic damage of
Congenital nephrotic syndrome, Finnish type, genetic damage of
Congenital nonhemolytic jaundice, genetic damage of
Congenital retinal telangiectasia, genetic damage of
Congenital short bowel, genetic damage of
Congenital short femur, genetic damage of
Congenital skeletal disorder, genetic damage of
Congenital skin disorder, genetic damage of
Congenital spherocytic anemia, genetic damage of
Congenital spherocytic hemolytic anemia, genetic damage of
Congenital stenosis of cervical medullary canal, genetic damage of
Congenital sucrose isomaltose malabsorption, genetic damage of
Congenital unilateral pulmonary hypoplasia, genetic damage of
Congenital vagal hyperreflexivity, genetic damage of
Congenital varicella syndrome, genetic damage of
Conn's syndrome, genetic damage of
Connective tissue dysplasia Spellacy type, genetic damage of
Connexin 26 anomaly, genetic damage of
Conotruncal heart malformations, genetic damage of
Conradi-Hünermann syndrome, genetic damage of
Constitutional growth delay, genetic damage of
Constrictive bronchiolitis, genetic damage of

Continuous muscle fiber activity hereditary, genetic damage of

Continuous spike-wave during slow sleep syndrome, genetic damage of

Contractural arachnodactyly, genetic damage of

Contractures ectodermal dysplasia cleft lip palate, genetic damage of

Contractures hyperkeratosis lethality, genetic damage of

Contractures of feet-muscle atrophy-oculomotor apraxia, genetic damage of

Conversion disorder, genetic damage of

Convulsions benign familial neonatal, genetic damage of

Convulsions benign familial neonatal dominant form, genetic damage of

Cooks syndrome, genetic damage of

Cooley's anemia, genetic damage of

Copper transport disease, genetic damage of

Coprastasophobia, genetic damage of (possible)

Coprophobia, genetic damage of (possible)

Coproporhyria, genetic damage of

Cor biloculare, genetic damage of

Cor triatriatum, genetic damage of

Cormier Rustin Munnich syndrome, genetic damage of

Corneal anesthesia deafness mental retardation, genetic damage of

Corneal cerebellar syndrome, genetic damage of

Corneal crystals myopathy neuropathy, genetic damage of

Corneal dystrophy, genetic damage of

Corneal dystrophy epithelial short stature, genetic damage of

Corneal dystrophy ichthyosis microcephaly mental retardation, genetic damage of

Corneal dystrophy perceptive deafness, genetic damage of

Corneal dystrophy pigmentary anomaly malabsorption, genetic damage of

Corneal endothelium dystrophy, genetic damage of

Cornelia de Lange syndrome, genetic damage of
Corneodermatoosseous syndrome, genetic damage of
Coronal dentin dysplasia, genetic damage of
Coronal synostosis syndactyly jejunal atresia, genetic damage of
Coronaro-cardiac fistula, genetic damage of
Coronary arteries congenital malformation, genetic damage of
Coronary artery aneurysm, genetic damage of
Corpus callosum agenesis, genetic damage of
Corpus callosum agenesis double urinary collecting, genetic damage of
Corpus callosum agenesis neuronopathy, genetic damage of
Corpus callosum agenesis of blepharophimosis Robin type, genetic damage of
Corpus callosum agenesis of with chorioretinal abnormalities, genetic damage of
Corpus callosum agenesis polysyndactyly, genetic damage of
Corpus callosum dysgenesis cleft spasm, genetic damage of
Corpus callosum dysgenesis hypopituitarism, genetic damage of
Corpus callosum dysgenesis X linked recessive, genetic damage of
Corrected transposition, genetic damage of
Corsello Opitz syndrome, genetic damage of
Cortada Koussef Matsumoto syndrome, genetic damage of
Cortes Lacassie syndrome, genetic damage of
Cortical blindness mental retardation polydactyly, genetic damage of
Cortical degeneration of the cerebellum parenchymatous, genetic damage of
Cortical hyperostosis syndactyly, genetic damage of
Corticobasal degeneration, genetic damage of
Costello syndrome, genetic damage of
Costocoracoid ligament congenitally short, genetic damage of
Costovertebral segmentation defect mesomelia, genetic damage of
Cote Adamopoulos Pantelakis syndrome, genetic damage of

Cote Katsantoni syndrome, genetic damage of
Cousin Walbraum Cegarra syndrome, genetic damage of
Covesdem syndrome, genetic damage of
Cowchock Wapner Kurtz syndrome, genetic damage of
Cowden's disease, genetic damage of
Coxoauricular syndrome, genetic damage of
Cramer Niederdellmann syndrome, genetic damage of
Cramp-fasciculations syndrome, genetic damage of
Crandall syndrome, genetic damage of
Crane Heise syndrome, genetic damage of
Cranio osteoarthropathy, genetic damage of
Cranioacrofacial syndrome, genetic damage of
Craniocerebellocardiac dysplasia, genetic damage of
Craniodiaphyseal dysplasia, genetic damage of
Craniodigital syndrome mental retardation, genetic damage of
Cranioectodermal dysplasia, genetic damage of
Craniofacial and osseous defects mental retardation, genetic damage of
Craniofacial and skeletal defects, genetic damage of
Craniofacial deafness hand syndrome, genetic damage of
Craniofacial dysostosis, genetic damage of
Craniofacial dysostosis arthrogryposis progeroid appearence, genetic damage of
Craniofacial dysynostosis, genetic damage of
Craniofaciocardioskeletal syndrome, genetic damage of
Craniofaciocervical osteoglyphic dysplasia, genetic damage of
Craniofrontonasal dysplasia, genetic damage of
Craniofrontonasal syndrome Teebi type, genetic damage of
Craniometaphyseal dysplasia, genetic damage of
Craniometaphyseal dysplasia dominant type, genetic damage of
Craniometaphyseal dysplasia recessive type, genetic damage of
Craniomicromelic syndrome, genetic damage of
Craniostenosis, genetic damage of

Craniostenosis cataract, genetic damage of
Craniostenosis with congenital heart disease mental retardation, genetic damage of
Craniosynostosis, genetic damage of
Craniosynostosis alopecia brain defect, genetic damage of
Craniosynostosis arthrogryposis cleft palate, genetic damage of
Craniosynostosis autosomal dominant, genetic damage of
Craniosynostosis brachydactyly, genetic damage of
Craniosynostosis cleft lip palate arthrogryposis, genetic damage of
Craniosynostosis contractures cleft, genetic damage of
Craniosynostosis Dandy Walker hydrocephalus, genetic damage of
Craniosynostosis exostoses nevus epibulbar dermoid, genetic damage of
Craniosynostosis fibular aplasia, genetic damage of
Craniosynostosis Fontaine type, genetic damage of
Craniosynostosis Maroteaux Fonfria type, genetic damage of
Craniosynostosis mental retardation clefting syndrome, genetic damage of
Craniosynostosis mental retardation heart defects, genetic damage of
Craniosynostosis Philadelphia type, genetic damage of
Craniosynostosis radial aplasia syndrome, genetic damage of
Craniosynostosis synostoses hypertensive nephropathy, genetic damage of
Craniosynostosis Warman type, genetic damage of
Craniotelencephalic dysplasia, genetic damage of
Crawfurd syndrome, genetic damage of
Creatine deficiency, genetic damage of
CREST syndrome, genetic damage of
Cretinism, genetic damage of
Cretinism athyreotic, genetic damage of
Cri du chat syndrome, genetic damage of
Crigler Najjar syndrome type I, genetic damage of

Crisponi syndrome, genetic damage of
Criss cross syndrome, genetic damage of
Criswick-Schepens syndrome, genetic damage of
Crohn's disease, genetic damage of
Crohn's disease of the esophagus, genetic damage of
Crome syndrome, genetic damage of
Cronkhite-Canada disease, genetic damage of
Crossed polydactyly type 1, genetic damage of
Crossed polysyndactyly, genetic damage of
Crouzon craniofacial dysostosis, genetic damage of
Crouzon disease, genetic damage of
Crow-Fukase syndrome, genetic damage of
Cryoglobulinemia, genetic damage of
Cryophobia, genetic damage of (possible)
Cryptogenic organized pneumopathy, genetic damage of
Cryptomicrotia brachydactyly syndrome, genetic damage of
Cryptomicrotia brachydactyly syndrome excess fingers, genetic damage of
Cryptophthalmos-syndactyly syndrome, genetic damage of
Cryptorchidism arachnodactyly mental retardation, genetic damage of
Cryroglobulinemia, genetic damage of
Crystal deposit disease, genetic damage of
Crystallophobia, genetic damage of (possible)
CTX, genetic damage of
Culler Jones syndrome, genetic damage of
Curly hair ankyloblepharon nail dysplasia syndrome, genetic damage of
Currarino triad, genetic damage of
Curry Hall syndrome, genetic damage of
Curth-Macklin type ichthyosis hystrix, genetic damage of
Curtis Rogers Stevenson syndrome, genetic damage of
Cushing syndrome, familial, genetic damage of

Cushing's symphalangism, genetic damage of
Cushing's syndrome, genetic damage of
Cutaneous lupus erythematosus, genetic damage of
Cutaneous photosensitivity colitis lethal, genetic damage of
Cutaneous T-cell lymphoma, genetic damage of
Cutaneous vascularitis, genetic damage of
Cutis Gyrata syndrome of Beare and Stevenson, genetic damage of
Cutis gyratum acanthosis nigricans craniosynostosis, genetic damage of
Cutis laxa, genetic damage of
Cutis laxa corneal clouding mental retardation, genetic damage of
Cutis laxa osteoporosis, genetic damage of
Cutis laxa with joint laxity and retarded development, genetic damage of
Cutis laxa, dominant type, genetic damage of
Cutis laxa, recessive type 1, genetic damage of
Cutis laxa, recessive type 2, genetic damage of
Cutis marmorata telangiectatica congenita, genetic damage of
Cutis marmorata telangiectatica congenita, genetic damage of
Cutis verticis gyrata, genetic damage of
Cutis verticis gyrata mental deficiency, genetic damage of
Cutis verticis gyrata thyroid aplasia mental retard, genetic damage of ation, genetic damage of
Cutler Bass Romshe syndrome, genetic damage of
Cyclic neutropenia, genetic damage of
Cyclic vomiting syndrome, genetic damage of
Cypress facial neuromusculoskeletal syndrome, genetic damage of
Cystathionine beta synthetase deficiency, genetic damage of
Cystic adenomatoid malformation of lung, genetic damage of
Cystic angiomatosis of bone, diffuse, genetic damage of
Cystic fibrosis, genetic damage of
Cystic fibrosis gastritis megaloblastic anemia, genetic damage of
Cystic hamartoma of lung and kidney, genetic damage of

Cystic hygroma, genetic damage of
Cystic hygroma lethal cleft palate, genetic damage of
Cystic medial necrosis of aorta, genetic damage of
Cystin transport, protein defect of, genetic damage of
Cystinosis, genetic damage of
Cystinuria, genetic damage of
Cystinuria-lysinuria, genetic damage of
Cytochrome C oxidase deficiency, genetic damage of
Cytomegalic inclusion disease, genetic damage of
Cytoplasmic body myopathy, genetic damage of
Czeizel Brooser syndrome, genetic damage of
Czeizel Losonci syndrome, genetic damage of
Czeizel syndrome, genetic damage of

D

D ercole syndrome, genetic damage of
D-glycerate dehydrogenase deficiency, genetic damage of
D-glycericacidemia, genetic damage of
D-minus hemolytic uremic syndrome (D-HUS), genetic damage of
D-plus hemolytic uremic syndrome (D+HUS), genetic damage of
Da Silva syndrome, genetic damage of
Daentl Towsend Siegel syndrome, genetic damage of
Dahlberg Borer Newcomer syndrome, genetic damage of
Daish Hardman Lamont syndrome, genetic damage of
Dandy Walker facial hemangioma, genetic damage of
Dandy Walker macrocephaly, genetic damage of
Dandy Walker malformation postaxial polydactyly, genetic damage of
Dandy Walker syndrome recessive form, genetic damage of
Dandy-Walker syndrome, genetic damage of
Daneman Davy Mancer syndrome, genetic damage of
Darier's disease, genetic damage of
Davenport Donlan syndrome, genetic damage of
David syndrome, genetic damage of
Davis Lafer syndrome, genetic damage of
De Barsy syndrome, genetic damage of
De Hauwere Leroy Adriaenssens syndrome, genetic damage of
De Morsier syndrome, genetic damage of
De Santis Cacchione syndrome, genetic damage of
Deaf blind hypopigmentation, genetic damage of
Deafness alopecia hypogonadism, genetic damage of
Deafness conductive ptosis skeletal anomalies, genetic damage of
Deafness conductive stapedial ear malformation facial palsy, genetic damage of
Deafness congenital onychodystrophy recessive, genetic damage of
Deafness craniofacial syndrome, genetic damage of

Deafness enamel hypoplasia nail defects, genetic damage of
Deafness epiphyseal dysplasia short stature, genetic damage of
Deafness goiter stippled epiphyses, genetic damage of
Deafness hyperuricemia neurologic ataxia, genetic damage of
Deafness hypogonadism syndrome, genetic damage of
Deafness hypospadias metacarpal and metatarsal syndrome, genetic damage of
Deafness mesenteric diverticula of small bowel neuropathy, genetic damage of
Deafness mixed with perilymphatic Gusher, X linked, genetic damage of
Deafness nephritis ano rectal malformation, genetic damage of
Deafness neurosensory pituitary dwarfism, genetic damage of
Deafness nonsyndromic, Connexin 26 linked, genetic damage of
Deafness oligodontia syndrome, genetic damage of
Deafness onychodystrophy dominant form, genetic damage of
Deafness peripheral neuropathy arterial disease, genetic damage of
Deafness progressive cataract autosomal dominant, genetic damage of
Deafness skeletal dysplasia lip granuloma, genetic damage of
Deafness symphalangism, genetic damage of
Deafness vitiligo achalasia, genetic damage of
Deafness white hair contractures papillomas, genetic damage of
Deafness X linked, DFN3, genetic damage of
Deafness, autosomal dominant nonsyndromic sensorineural, genetic damage of
Deafness, isolated, due to mitochondrial transmission, genetic damage of
Deafness, neurosensory nonsyndromic recessive, DFN, genetic damage of
Deafness, X linked, DFN, genetic damage of
Deal Barratt Dillon syndrome, genetic damage of
Deciduous skin, genetic damage of

Decompensated phoria, genetic damage of
Defect in synthesis of adenosylcobalamin, genetic damage of
Defective apolipoprotein B-100, genetic damage of
Defective expression of HLA class 2, genetic damage of
Degenerative motor system disease, genetic damage of
Degenerative optic myopathy, genetic damage of
Degos 'en cocarde' erythrokeratoderma, genetic damage of
Degos disease, genetic damage of
Dehydratase deficiency, genetic damage of
Dehydrated hereditary stomatocytosis, genetic damage of
Deipnophobia, genetic damage of (possible)
Dejerine-Sottas disease, genetic damage of
Delayed membranous cranial ossification, genetic damage of
Delayed speech facial asymmetry strabismus ear lobe creases, genetic damage of
Deletion 10p, genetic damage of
Deletion 10q, genetic damage of
Deletion 11p, genetic damage of
Deletion 11p 11p12, genetic damage of
Deletion 11p13, genetic damage of
Deletion 11q partial, genetic damage of
Deletion 12p12 p11, genetic damage of
Deletion 12p13, genetic damage of
Deletion 13q, genetic damage of
Deletion 13q14, genetic damage of
Deletion 13q22, genetic damage of
Deletion 13q32, genetic damage of
Deletion 14q partial duplication 14p partial, genetic damage of
Deletion 14q11, genetic damage of
Deletion 14q31, genetic damage of
Deletion 14qter, genetic damage of
Deletion 15q1, genetic damage of
Deletion 15q25, genetic damage of

Deletion 17q23 q24, genetic damage of
Deletion 18p, genetic damage of
Deletion 18q, genetic damage of
Deletion 18q23, genetic damage of
Deletion 1p, genetic damage of
Deletion 1p22 p13, genetic damage of
Deletion 1p31 p22, genetic damage of
Deletion 1p32, genetic damage of
Deletion 1p34 p32, genetic damage of
Deletion 1q21 q25, genetic damage of
Deletion 1q25 q32, genetic damage of
Deletion 1q32 q42, genetic damage of
Deletion 1q4, genetic damage of
Deletion 20p, genetic damage of
Deletion 21q22, genetic damage of
Deletion 2p22, genetic damage of
Deletion 2pter p24, genetic damage of
Deletion 2q, genetic damage of
Deletion 2q duplication 1p, genetic damage of
Deletion 2q24, genetic damage of
Deletion 3p, genetic damage of
Deletion 3p14 p11, genetic damage of
Deletion 3p25, genetic damage of
Deletion 3q13, genetic damage of
Deletion 3q21 23, genetic damage of
Deletion 3q27, genetic damage of
Deletion 4p, genetic damage of
Deletion 4p14 p16, genetic damage of
Deletion 4q, genetic damage of
Deletion 4q32, genetic damage of
Deletion 5q35, genetic damage of
Deletion 6p23, genetic damage of
Deletion 6q, genetic damage of

Deletion 6q1, genetic damage of
Deletion 6q13 q15, genetic damage of
Deletion 6q16 q21, genetic damage of
Deletion 6q2, genetic damage of
Deletion 7, genetic damage of
Deletion 7q2, genetic damage of
Deletion 7q21, genetic damage of
Deletion 7q3, genetic damage of
Deletion 8p, genetic damage of
Deletion 8p23 1, genetic damage of
Deletion 8q, genetic damage of
Deletion 8q12 21, genetic damage of
Deletion 8q21 q22, genetic damage of
Deletion 9p, genetic damage of
Deletion Xp22 pter, genetic damage of
Deletion Xq28, genetic damage of
Delleman oorthuys syndrome, genetic damage of
Delta-1-pyrroline-5-carboxylate dehydrogenase deficiency, genetic damage of
Delta-sarcoglycanopathy, genetic damage of
Dementia progressive lipomembranous polycysta, genetic damage of
Dementophobia, genetic damage of (possible)
Demonophobia, genetic damage of (possible)
Demyelinating diseases, genetic damage of
Dendrophobia, genetic damage of (possible)
Dennis Cohen syndrome, genetic damage of
Dennis Fairhurst Moore syndrome, genetic damage of
Dent disease, genetic damage of
Dental aberrations steroid dehydrogenase deficiency, genetic damage of
Dental tissue neoplasm, genetic damage of
Dentatorubral pallidoluysian atrophy, genetic damage of

Dentin dysplasia sclerotic bones, genetic damage of
Dentin dysplasia, coronal, genetic damage of
Dentin dysplasia, radicular, genetic damage of
Dentinogenesis imperfecta, genetic damage of
Dentophobia, genetic damage of (possible)
Depersonalization disorder, genetic damage of
Der kaloustian Jarudi Khoury syndrome, genetic damage of
Der Kaloustian McIntosh Silver syndrome, genetic damage of
Dercum disease, genetic damage of
Dermatitis herpetiformis, genetic damage of
Dermatocardioskeletal syndrome Boronne type, genetic damage of
Dermatographic uticaria, genetic damage of
Dermatoleukodystrophy, genetic damage of
Dermatomyositis, genetic damage of
Dermatoosteolysis Kirghizian type, genetic damage of
Dermatophobia, genetic damage of (possible)
Dermochondrocorneal dystrophy of François, genetic damage of
Dermoodontodysplasia, genetic damage of
Dermopathy restrictive lethal, genetic damage of
Desbuquois syndrome, genetic damage of
Desmin related myopathy, genetic damage of
Desmoid disease, genetic damage of
Desmoid tumor, genetic damage of
Desmoplastic small cell tumor, genetic damage of
Developmental delay hypotonia extremities hypertrophy, genetic damage of
Developmental dysphasia familial, genetic damage of
Devic syndrome, genetic damage of
Devriendt Legius Fryns syndrome, genetic damage of
Devriendt Vandenberghe Fryns syndrome, genetic damage of
Dexamethasone sensitive hypertension, genetic damage of
Dextrocardia, genetic damage of
Dextrocardia with situs inversus, genetic damage of

Dextrocardia-bronchiectasis-sinusitis, genetic damage of

Diabetes hypogonadism deafness mental retardation, genetic damage of

Diabetes insipidus, diabetes mellitus, optic atrophy, genetic damage of

Diabetes insipidus, nephrogenic type 1, genetic damage of

Diabetes insipidus, nephrogenic type 2, genetic damage of

Diabetes insipidus, nephrogenic type 3, genetic damage of

Diabetes insipidus, nephrogenic, dominant type, genetic damage of

Diabetes insipidus, nephrogenic, recessive type, genetic damage of

Diabetes mellitus, transient neonatal, genetic damage of

Diabetes persistent mullerian ducts, genetic damage of

Diabetes, insulin dependent, genetic damage of

Diabetic angiopathy, genetic damage of

Diabetic embryopathy, genetic damage of

Diabetic nephropathy, genetic damage of

Diabetic neuropathy, genetic damage of

Diamond Blackfan disease, genetic damage of

Diaphragmatic agenesia, genetic damage of

Diaphragmatic agenesis radial aplasia omphalocele, genetic damage of

Diaphragmatic defect limb deficiency skull defect, genetic damage of

Diaphragmatic hernia abnormal face limb, genetic damage of

Diaphragmatic hernia exomphalos corpus callosum agenesis, genetic damage of

Diaphragmatic hernia upper limb defects, genetic damage of

Diaphragmatic hernia, congenital, genetic damage of

Diarrhea chronic with villous atrophy, genetic damage of (possible)

Diarrhea polyendocrinopathy infections X linked, genetic damage of

Diastematomyelia, genetic damage of
Diastrophic dwarfism, genetic damage of
Diastrophic dysplasia, genetic damage of
Dibasic aminoaciduria 2, genetic damage of
Dibasic aminoaciduria type 1, genetic damage of
Dicarboxylicaminoaciduria, genetic damage of
Die Smulders Droog Van Dijk syndrome, genetic damage of
Die Smulders Vles Fryns syndrome, genetic damage of
Diencephalic syndrome, genetic damage of
Dieterich's disease, genetic damage of
Diffuse idiopathic skeletal hyperostosis, genetic damage of
Diffuse leiomyomatosis with Alport syndrome, genetic damage of
Diffuse neonatal haemangiomatosis, genetic damage of
Diffuse palmoplantar keratoderma Bothnian type, genetic damage of
DiGeorge syndrome, genetic damage of
Digestive duplication, genetic damage of
Digitorenocerebral syndrome, genetic damage of
Dihydropteridine reductase deficiency, genetic damage of
Dihydropyrimidine dehydrogenase deficiency, genetic damage of
Dilated cardiomyopathy, genetic damage of
Dimitri Sturge Weber syndrome, genetic damage of
Dincsoy Salih Patel syndrome, genetic damage of
Dinno Shearer Weisskopf syndrome, genetic damage of
Dinophobia, genetic damage of (possible)
Diomedi Bernardi Placidi syndrome, genetic damage of
Dionisi Vici Sabetta Gambarara syndrome, genetic damage of
Diphallia, genetic damage of
Diphallus rachischisis imperforate anus, genetic damage of
Diphosphoglycerate mutase deficiency of erythrocyte, genetic damage of
Diplophobia, genetic damage of (possible)
Dipsophobia, genetic damage of (possible)

Discoid lupus erythematosus, genetic damage of
Dislocation of the hip dysmorphism, genetic damage of
Disomy 1q12 q21, genetic damage of
Disomy 9q21, genetic damage of
Disorder in the hormonal synthesis with or without goiter, genetic damage of
Disorganization syndrome, genetic damage of
Dissecting cellulitis of the scalp, genetic damage of
Dissociative hysteria, genetic damage of
Distal arthrogryposis Moore Weaver type, genetic damage of
Distal myopathy, genetic damage of
Distal myopathy Markesbery-Griggs type, genetic damage of
Distal myopathy with vocal cord weakness, genetic damage of
Distal myopathy, Nonaka type, genetic damage of
Distal primary acidosis, familial, genetic damage of
Distichiasis heart congenital anomalies, genetic damage of
DK phocomelia syndrome, genetic damage of
Dobrow syndrome, genetic damage of
Dominant cleft palate, genetic damage of
Dominant ichthyosis vulgaris, genetic damage of
Dominant zonular cataract, genetic damage of
Donnai Barrow syndrome, genetic damage of
Donohue syndrome, genetic damage of
Door syndrome, genetic damage of
Dopa-responsive dystonia, genetic damage of
Dopamine beta-hydroxylase deficiency, genetic damage of
Doraphobia, genetic damage of (possible)
Double cortex, genetic damage of
Double discordia, genetic damage of
Double fingernail of fifth finger, genetic damage of
Double outlet left ventricle, genetic damage of
Double outlet right ventricle, genetic damage of
Double tachycardia induced by catecholamines, genetic damage of

Double uterus-hemivagina-renal agenesis, genetic damage of
Downs syndrome, genetic damage of
Doxorubicin-induced cardiomyopathy, genetic damage of
Doyne honeycomb retinal dystrophy, genetic damage of
Drachtman Weinblatt Sitarz syndrome, genetic damage of
Duane anomaly mental retardation, genetic damage of
Duane syndrome, genetic damage of
Dubin-Johnson syndrome, genetic damage of
Dubowitz syndrome, genetic damage of
Duchenne muscular dystrophy, genetic damage of
Ductular hepatic hypoplasia, genetic damage of
Duhring Brocq disease, genetic damage of
Duhring's disease, genetic damage of
Duker Weiss Siber syndrome, genetic damage of
Duodenal atresia, genetic damage of
Duodenal atresia tetralogy of Fallot, genetic damage of
Duplication 10p, genetic damage of
Duplication 10pter p13, genetic damage of
Duplication 10q partial, genetic damage of
Duplication 11q, genetic damage of
Duplication 11q23, genetic damage of
Duplication 12p, genetic damage of
Duplication 12q, genetic damage of
Duplication 13, genetic damage of
Duplication 13p, genetic damage of
Duplication 14q partial deletion 14p partial, genetic damage of
Duplication 14q prox, genetic damage of
Duplication 14q ter, genetic damage of
Duplication 15q, genetic damage of
Duplication 16p, genetic damage of
Duplication 16q, genetic damage of
Duplication 17p, genetic damage of
Duplication 17p11.2, genetic damage of

Duplication 18, genetic damage of
Duplication 18p, genetic damage of
Duplication 18q, genetic damage of
Duplication 19q, genetic damage of
Duplication 1p21 p32, genetic damage of
Duplication 1q12 q21, genetic damage of
Duplication 1q32 qter, genetic damage of
Duplication 1q42 qter, genetic damage of
Duplication 1q42.11 q42.12, genetic damage of
Duplication 20p, genetic damage of
Duplication 22, genetic damage of
Duplication 22q11 q13, genetic damage of
Duplication 2p, genetic damage of
Duplication 2p13 p21, genetic damage of
Duplication 2pter p24, genetic damage of
Duplication 2q, genetic damage of
Duplication 2q37, genetic damage of
Duplication 3p, genetic damage of
Duplication 3p25, genetic damage of
Duplication 3q, genetic damage of
Duplication 3q13.2 q25, genetic damage of
Duplication 4p, genetic damage of
Duplication 4q, genetic damage of
Duplication 4q21, genetic damage of
Duplication 4q25 qter, genetic damage of
Duplication 5pter p13.3, genetic damage of
Duplication 5q, genetic damage of
Duplication 6p, genetic damage of
Duplication 6q, genetic damage of
Duplication 7p, genetic damage of
Duplication 7p13 p12.2, genetic damage of
Duplication 7q, genetic damage of
Duplication 8p, genetic damage of

Duplication 8q, genetic damage of
Duplication 9p partial, genetic damage of
Duplication 9q21, genetic damage of
Duplication 9q32, genetic damage of
Duplication of leg mirror foot, genetic damage of
Duplication of the thumb unilateral biphalangeal, genetic damage of
Duplication of urethra, genetic damage of
Duplication Xp3, genetic damage of
Duplication Xpter Xq13, genetic damage of
Duplication Xq, genetic damage of
Duplication Xq13 1 q21 1, genetic damage of
Duplication Xq25, genetic damage of
Dupont Sellier Chochillon syndrome, genetic damage of
Dupuytren's contracture, genetic damage of
Dwarfism, genetic damage of
Dwarfism bluish sclerae, genetic damage of
Dwarfism deafness retinitis pigmentosa, genetic damage of
Dwarfism lethal type advanced bone age, genetic damage of
Dwarfism mental retardation eye abnormality, genetic damage of
Dwarfism short limb absent fibulas very short digits, genetic damage of
Dwarfism stiff joint ocular abnormalities, genetic damage of
Dwarfism syndesmodysplasic, genetic damage of
Dwarfism tall vertebrae, genetic damage of
Dwarfism thanatophoric, genetic damage of
Dwarfism thin bones multiple fractures, genetic damage of
Dyggve-Melchior-Clausen syndrome, genetic damage of
Dykes Markes Harper syndrome, genetic damage of
Dysautonomia (does not have to be familial), genetic damage of
Dyschondrosteosis, genetic damage of
Dyschondrosteosis nephritis, genetic damage of
Dyschromatosis universalis, genetic damage of

Dysencephalia splachnocystica or Meckel Gruber, genetic damage of

Dysequilibrium syndrome, genetic damage of

Dyserythropoietic anemia, congenital, genetic damage of

Dyserythropoietic anemia, congenital type 1, genetic damage of

Dyserythropoietic anemia, congenital type 2, genetic damage of

Dyserythropoietic anemia, congenital type 3, genetic damage of

Dysferlinopathy, genetic damage of

Dysfibrinogenemia, familial, genetic damage of

Dysgerminoma, genetic damage of

Dysharmonic skeletal maturation muscular fiber disproportion, genetic damage of

Dyskeratosis congenital syndrome, genetic damage of

Dyskeratosis congenita of Zinsser Cole Engman, genetic damage of

Dyskeratosis follicularis, genetic damage of

Dysmorphism abnormal vocalization mental retardation, genetic damage of

Dysmorphism cleft palate loose skin, genetic damage of

Dysmorphism multiple structural anomalies, genetic damage of

Dysmorphophobia, genetic damage of (possible)

Dysosteosclerosis, genetic damage of

Dysostosis, genetic damage of

Dysostosis acral with facial and genital abnormalities, genetic damage of

Dysostosis acrofacial postaxial, genetic damage of

Dysostosis peripheral, genetic damage of

Dysostosis Stanescu type, genetic damage of

Dysphasic dementia, hereditary, genetic damage of

Dysphonia, chronic spasmodic, genetic damage of

Dysplasia, genetic damage of

Dysplasia epiphysealis hemimelica, genetic damage of

Dysplasia faciogenital, genetic damage of

Dysplasia olfactogenitalis of de Morsier, genetic damage of
Dysplastic cortical hyperostosis, genetic damage of
Dysplastic nevus syndrome, genetic damage of
Dysprothrombinemia, genetic damage of
Dysraphism cleft lip palate limb reduction defects, genetic damage of
Dyssegmental dysplasia glaucoma, genetic damage of
Dyssegmental dysplasia Silverman Handmaker type, genetic damage of
Dysthymia, genetic damage of
Dystonia, genetic damage of
Dystonia musculorum deformans, genetic damage of
Dystonia musculorum deformans type 1, genetic damage of
Dystonia musculorum deformans type 2, genetic damage of
Dystonia progressive with diurnal variation, genetic damage of
Dystrophic epidermolysis bullosa, genetic damage of
Dystrophinopathy, genetic damage of
Dystrophy, myotonic, genetic damage of
Dystychiphobia, genetic damage of (possible)

E

EAF, genetic damage of

Eales disease, genetic damage of

Ear patella short stature syndrome, genetic damage of

Earlobes thickened conductive deafness, genetic damage of

Early infantile autism, genetic damage of

Eaton-Lambert syndrome, genetic damage of

Ebstein's anomaly, genetic damage of

Eccentrochondrodysplasia, genetic damage of

Eccrine acrospiroma, genetic damage of

Eclampsia, genetic damage of

Ecp syndrome, genetic damage of

Ectodermal dysplasia, genetic damage of

Ectodermal dysplasia absent dermatoglyphics, genetic damage of

Ectodermal dysplasia adrenal cyst, genetic damage of

Ectodermal dysplasia alopecia preaxial polydactyly, genetic damage of

Ectodermal dysplasia anhidrotic, genetic damage of

Ectodermal dysplasia arthrogryposis diabetes mellitus, genetic damage of

Ectodermal dysplasia Bartalos type, genetic damage of

Ectodermal dysplasia Berlin type, genetic damage of

Ectodermal dysplasia blindness, genetic damage of

Ectodermal dysplasia ectrodactyly macular dystrophy, genetic damage of

Ectodermal dysplasia hypohidrotic autosomal dominant, genetic damage of

Ectodermal dysplasia hypohidrotic hypothyroidism ciliary diskinesia, genetic damage of

Ectodermal dysplasia Margarita type, genetic damage of

Ectodermal dysplasia mental retardation CNS malformation, genetic damage of

Ectodermal dysplasia mental retardation syndactyly, genetic dam-

age of

Ectodermal dysplasia neurosensory deafness, genetic damage of

Ectodermal dysplasia osteosclerosis, genetic damage of

Ectodermal dysplasia tricho odonto onychial type, genetic damage of

Ectodermal dysplasia, hydrotic, genetic damage of

Ectodermal dysplasia, hypohidrotic, autosomal recessive, genetic damage of

Ectodermal dysplasias, genetic damage of

Ectodermic dysplasia anhidrotic cleft lip, genetic damage of

Ectopia lentis chorioretinal dystrophy myopia, genetic damage of

Ectopia lentis isolated, genetic damage of

Ectopic coarctation, genetic damage of

Ectopic ossification familial type, genetic damage of

Ectopic pregnancy, genetic damage of

Ectrodactyly, genetic damage of

Ectrodactyly cardiopathy dysmorphism, genetic damage of

Ectrodactyly cleft palate syndrome, genetic damage of

Ectrodactyly diaphragmatic hernia corpus callosum, genetic damage of

Ectrodactyly dominant form, genetic damage of

Ectrodactyly ectrodermal dysplasia, genetic damage of

Ectrodactyly polydactyly, genetic damage of

Ectrodactyly recessive form, genetic damage of

Ectrodactyly spina bifida cardiopathy, genetic damage of

Ectrodactyly-ectodermal dysplasia-cleft lip/cleft palate, genetic damage of

Ectropion inferior cleft lip and or palate, genetic damage of

Edinburgh malformation syndrome, genetic damage of

Edwards Patton Dilly syndrome, genetic damage of

Edwards syndrome, genetic damage of

Eec syndrome, genetic damage of

Eec syndrome without cleft lip palate, genetic damage of

EEM syndrome, genetic damage of
Ehlers-Danlos syndrome, genetic damage of
Ehlers-Danlos syndrome type 1, genetic damage of
Ehlers-Danlos syndrome type 2, genetic damage of
Ehlers-Danlos syndrome type 3, genetic damage of
Ehlers-Danlos syndrome type 4, autosomal dominant, genetic damage of
Ehlers-Danlos syndrome type 6, genetic damage of
Ehlers-Danlos syndrome type 7A, genetic damage of
Ehlers-Danlos syndrome type 7B, genetic damage of
Ehlers-Danlos syndrome type 7C, genetic damage of
Ehlers-Danlos syndrome, arthrochalasic type, genetic damage of
Ehlers-Danlos syndrome, classic type, genetic damage of
Ehlers-Danlos syndrome, dermatosparaxis type, genetic damage of
Ehlers-Danlos syndrome, hypermobile type, genetic damage of
Ehlers-Danlos syndrome, vascular type, genetic damage of
Eijkman's syndrome, genetic damage of
Eisenmenger syndrome, genetic damage of
Eisoptrophobia, genetic damage of (possible)
Elattoproteus in context of NF, genetic damage of
Elective mutism, genetic damage of
Electron transfer flavoprotein, deficiency of, genetic damage of
Electrophobia, genetic damage of (possible)
Elejalde syndrome, genetic damage of
Elephant man in context of NF, genetic damage of
Elephantiasis, genetic damage of
Elliott Ludman Teebi syndrome, genetic damage of
Ellis Yale Winter syndrome, genetic damage of
Ellis-Van Creveld syndrome, genetic damage of
Emerinopathy, genetic damage of
Emery Nelson syndrome, genetic damage of
Emery-Dreifuss muscular dystrophy, genetic damage of
Emery-Dreifuss muscular dystrophy, dominant type, genetic dam-

age of

Emery-Dreifuss muscular dystrophy, X linked, genetic damage of

Emetophobia, genetic damage of (possible)

Emphysema, genetic damage of

Emphysema, congenital lobar, genetic damage of

Emphysema-penoscrotal web-deafness-mental retardation, genetic damage of

Empty sella syndrome, genetic damage of

Enamel hypoplasia cataract hydrocephaly, genetic damage of

Enamel renal syndrome, genetic damage of

Encephalo cranio cutaneous lipomatosis, genetic damage of

Encephalocele, genetic damage of

Encephalocele anencephaly, genetic damage of

Encephalocele anterior, genetic damage of

Encephalocele frontal, genetic damage of

Encephalomyelitis, genetic damage of (possible)

Encephalomyelitis, myalgic, genetic damage of (possible)

Encephalopathy intracerebral calcification retinal, genetic damage of

Encephalopathy progressive optic atrophy, genetic damage of

Encephalopathy subacute spongiform, Gerstmann-Stra, genetic damage of

Encephalopathy-basal ganglia-calcification, genetic damage of

Encephalophathy recurrent of childhood, genetic damage of

Enchondromatosis (benign), genetic damage of

Enchondromatosis dwarfism deafness, genetic damage of

Endocardial fibroelastosis, genetic damage of

Endocrinopathy, genetic damage of

Endometrial stromal sarcoma, genetic damage of

Endometriosis, genetic damage of

Endomyocardial fibroelastosis, genetic damage of

Endomyocardial fibrosis, genetic damage of

Enetophobia, genetic damage of (possible)

Eng Strom syndrome, genetic damage of
Engelhard Yatziv syndrome, genetic damage of
Englemann disease, genetic damage of
Enochlophobia, genetic damage of (possible)
Enolase deficiency, genetic damage of
Enolase deficiency type 1, genetic damage of
Enolase deficiency type 2, genetic damage of
Enolase deficiency type 3, genetic damage of
Enolase deficiency type 4, genetic damage of
Enteropathica, genetic damage of
Eosinophilic cystitis, genetic damage of
Eosinophilic fasciitis, genetic damage of
Eosinophilic gastroenteritis, genetic damage of
Eosinophilic granuloma, genetic damage of
Eosinophilic lymphogranuloma, genetic damage of
Eosinophilic synovitis, genetic damage of
Eosophobia, genetic damage of (possible)
Ependymoblastoma, genetic damage of
Ependymoma, genetic damage of
Epidermal nevus syndrome, genetic damage of
Epidermal nevus vitamin D resistant rickets, genetic damage of
Epidermodysplasia verruciformis, genetic damage of
Epidermoid carcinoma, genetic damage of
Epidermolysa bullosa simplex and limb girdle muscular dystrophy, genetic damage of
Epidermolysis bullosa, genetic damage of
Epidermolysis bullosa acquisita, genetic damage of
Epidermolysis bullosa dystrophica, Bart type, genetic damage of
Epidermolysis bullosa dystrophica, dominant type, genetic damage of
Epidermolysis bullosa dystrophica, Hallopeau-Sieme, genetic damage of
Epidermolysis bullosa herpetiformis, Dowling-Meara, genetic

damage of
Epidermolysis bullosa intraepidermic, genetic damage of
Epidermolysis bullosa inversa dystrophica, genetic damage of
Epidermolysis bullosa of hands and feet, genetic damage of
Epidermolysis bullosa simplex with anodontia, hair, genetic damage of
Epidermolysis bullosa simplex, Cockayne-Touraine type, genetic damage of
Epidermolysis bullosa simplex, Koebner type, genetic damage of
Epidermolysis bullosa simplex, Ogna type, genetic damage of
Epidermolysis bullosa simplex, Weber-Cockayne type, genetic damage of
Epidermolysis bullosa, dermolytic, genetic damage of
Epidermolysis bullosa, generalized atrophic benign, genetic damage of
Epidermolysis bullosa, junctional, genetic damage of
Epidermolysis bullosa, junctional, Herlitz-Pearson, genetic damage of
Epidermolysis bullosa, junctional, with pyloric atrophy, genetic damage of
Epidermolysis bullosa, pretibial, genetic damage of
Epidermolytic hyperkeratosis, genetic damage of
Epidermolytic palmoplantar keratoderma Vorner type, genetic damage of
Epilepsy, genetic damage of
Epilepsy benign neonatal dominant form, genetic damage of
Epilepsy benign neonatal recessive form, genetic damage of
Epilepsy juvenile absence, genetic damage of
Epilepsy mental deterioration Finnish type, genetic damage of
Epilepsy microcephaly skeletal dysplasia, genetic damage of
Epilepsy occipital calcifications, genetic damage of
Epilepsy progressive myoclonic type 2, genetic damage of
Epilepsy telangiectasia, genetic damage of

Epilepsy with myoclono-astatic crisis, genetic damage of
Epilepsy, benign occipital, genetic damage of
Epilepsy, myoclonic progressive familial, genetic damage of
Epilepsy, nocturnal, frontal lobe type, genetic damage of
Epilepsy, partial, familial, genetic damage of
Epilepsy, progressive myoclonic type 1, genetic damage of
Epimerase deficiency, genetic damage of
Epimetaphyseal dysplasia cataract, genetic damage of
Epimetaphyseal skeletal dysplasia, genetic damage of
Epiphyseal dysplasia dysmorphism camptodactyly, genetic damage of
Epiphyseal dysplasia hearing loss dysmorphism, genetic damage of
Epiphyseal dysplasia multiple, genetic damage of
Epiphyseal stippling syndrome osteoclastic hyperplasia, genetic damage of
Epiphysealis hemimelica dysplasia, genetic damage of
Epistaxiophobia, genetic damage of (possible)
Epithelial-myoepithelial carcinoma, genetic damage of
Epitheliopathy (APMPPE), genetic damage of
Epitheliopathy, acute posterior multifocal placoid, genetic damage of
EPP (erythropoietic protoporphyria), genetic damage of
Epstein syndrome, genetic damage of
Equinophobia, genetic damage of (possible)
Erb's palsy, genetic damage of, genetic damage of
Erb-Duchenne palsy, genetic damage of
Erdheim disease, genetic damage of
Erdheim-Chester syndrome, genetic damage of
Ereuthrophobia, genetic damage of (possible)
Ergophobia, genetic damage of (possible)
Eronen Somer Gustafsson syndrome, genetic damage of
Erosive pustular dermatosis of the scalp, genetic damage of
Erythema multiforme, genetic damage of

Erythermalgia, genetic damage of
Erythroblastopenia, genetic damage of
Erythroderma desquamativa of Leiner, genetic damage of
Erythroderma lethal congenital, genetic damage of
Erythrokeratodermia ataxia, genetic damage of
Erythrokeratodermia progressive symmetrica ichthyosis, genetic damage of
Erythrokeratodermia symmetrica progressiva, genetic damage of
Erythrokeratodermia variabilis ichthyosis, genetic damage of
Erythrokeratodermia variabilis, Mendes da Costa type, genetic damage of
Erythrokeratodermia with ataxia, genetic damage of
Erythrokeratolysis hiemalis ichthyosis, genetic damage of
Erythromelalgia, genetic damage of
Erythroplakia, genetic damage of
Erythropoietic protoporphyria, genetic damage of
Escher Hirt syndrome, genetic damage of
Esophageal atresia, genetic damage of
Esophageal atresia associated anomalies, genetic damage of
Esophageal atresia coloboma talipes, genetic damage of
Esophageal disorder, genetic damage of
Esophageal duodenal atresia abnormalities of hands, genetic damage of
Esophageal neoplasm, genetic damage of
Esophageal varices, genetic damage of
Esotropia, genetic damage of
Essential hypertension, genetic damage of
Essential iris atrophy, genetic damage of
Essential mixed cryoglobulinemia, genetic damage of
Essential thrombocytopenia, genetic damage of
Essential thrombocytosis, genetic damage of
Esthesioneuroblastoma, genetic damage of
Ethylmalonic aciduria, genetic damage of

Euhidrotic ectodermal dysplasia, genetic damage of
Eunuchoidism familial, genetic damage of
Euphobia, genetic damage of (possible)
Evan's syndrome, genetic damage of
Ewing's sarcoma, genetic damage of
Exencephaly, genetic damage of
Exfoliative dermatitis, genetic damage of
Exner syndrome, genetic damage of
Exomphalos-macroglossia-gigantism syndrome, genetic damage of
Exostoses, genetic damage of
Exostoses anetodermia brachydactyly type E, genetic damage of
Exostoses, multiple, genetic damage of
Exostoses, multiple, type 1, genetic damage of
Exostoses, multiple, type 2, genetic damage of
Exostoses, multiple, type 3, genetic damage of
Exstrophy of the bladder, genetic damage of
Exstrophy of the bladder-epispadias, genetic damage of
Exsudative retinopathy familial, autosomal dominant, genetic damage of
Exsudative retinopathy familial, autosomal recessive, genetic damage of
Exsudative retinopathy familial, X linked, recessive, genetic damage of
Exsudative retinopathy, familial, genetic damage of
Extrapyramidal disorder, genetic damage of
Extrasystoles short stature hyperpigmentation microcephaly, genetic damage of
Eye defects arachnodactyly cardiopathy, genetic damage of
Eyebrows and eyelashes absence mental retardation, genetic damage of
Eyebrows duplication syndactyly, genetic damage of

F

Fabry's disease, genetic damage of
Faces syndrome, genetic damage of
Facial asymetry temporal seizures, genetic damage of
Facial clefting corpus callosum agenesis, genetic damage of
Facial dysmorphism macrocephaly myopia Dandy Walker type, genetic damage of
Facial dysmorphism shawl scrotum joint laxity syndrome, genetic damage of
Facial paralysis, genetic damage of (possible)
Facies unusual arthrogryposis advanced skeletal malformations, genetic damage of
Facio digito genital syndrome recessive form, genetic damage of
Facio skeletal genital syndrome Rippberger type, genetic damage of
Facio thoraco genital syndrome, genetic damage of
Faciocardiomelic dysplasia lethal, genetic damage of
Faciocardiorenal syndrome, genetic damage of
Faciodigitogenital syndrome, genetic damage of
Faciooculoacousticorenal syndrome, genetic damage of
Facioscapulohumeral muscular dystrophy, genetic damage of
Faciothoracoskeletal syndrome, genetic damage of
Factor II deficiency, genetic damage of
Factor II deficiency, genetic damage of
Factor IX deficiency, genetic damage of
Factor V deficiency, genetic damage of
Factor V Leiden mutation, genetic damage of
Factor VII deficiency, genetic damage of
Factor VIII deficiency, genetic damage of
Factor X deficiency, genetic damage of
Factor X deficiency, congenital, genetic damage of
Factor XI deficiency, congenital, genetic damage of
Factor XII deficiency, genetic damage of

Factor XIII deficiency, genetic damage of
Factor XIII deficiency, congenital, genetic damage of
Fahr's disease, genetic damage of
Fairbank disease, genetic damage of
Fallot complex mental growth retardation, genetic damage of
Fallot tetralogy, genetic damage of
Familial adenomatous polyposis, genetic damage of
Familial amyloid polyneuropathy, genetic damage of
Familial aortic dissection, genetic damage of
Familial band heterotopia, genetic damage of
Familial chondrocalcinosis, genetic damage of
Familial deafness, genetic damage of (possible)
Familial dysautonomia, genetic damage of
Familial emphysema, genetic damage of
Familial erythrophagocytic lymphohistiocytosis, genetic damage of
Familial hyperchylomicronemia, genetic damage of
Familial hyperlipoproteinemia, genetic damage of
Familial hyperlipoproteinemia type I, genetic damage of
Familial hyperlipoproteinemia type III, genetic damage of
Familial hyperlipoproteinemia type IV, genetic damage of
Familial hypersensitivity pneumonitis, genetic damage of
Familial hypertension, genetic damage of (possible)
Familial hypopituitarism, genetic damage of
Familial hypothyroidism, genetic damage of
Familial intestinal polyatresia syndrome, genetic damage of
Familial nasal acilia, genetic damage of
Familial non-immune hyperthyroidism, genetic damage of
Familial opposable triphalangeal thumbs duplication, genetic damage of
Familial partial epilepsy with variable focus, genetic damage of
Familial periodic paralysis, genetic damage of
Familial polyposis, genetic damage of
Familial porencephaly, genetic damage of

Familial supernumerary nipples, genetic damage of
Familial symmetric lipomatosis, genetic damage of
Familial thyroglossal duct cyst, genetic damage of
Familial Treacher Collins syndrome, genetic damage of
Familial veinous malformations, genetic damage of
Familial ventricular tachycardia, genetic damage of
Familial visceral myopathy, genetic damage of
Fanconi anemia type 1, genetic damage of
Fanconi anemia type 2, genetic damage of
Fanconi anemia type 3, genetic damage of
Fanconi Bickel syndrome, genetic damage of
Fanconi ichthyosis dysmorphism, genetic damage of
Fanconi like syndrome, genetic damage of
Fanconi pancytopenia, genetic damage of
Fanconi syndrome, renal, with nephrocalcinosis and renal stones, genetic damage of
Fanconi's anemia, genetic damage of
Fara Chlupackova syndrome, genetic damage of
Farber's disease, genetic damage of
Fas deficiency, genetic damage of
Fatal familial insomnia, genetic damage of
Fatty Liver, genetic damage of
Faulk Epstein Jones syndrome, genetic damage of
Faye Petersen Ward Carey syndrome, genetic damage of
Fazio Londe syndrome, genetic damage of
Fealty syndrome, genetic damage of
Febrile seizure, genetic damage of
Fechtner syndrome, genetic damage of
Feigenbaum Bergeron Richardson syndrome, genetic damage of
Feigenbaum Bergeron syndrome, genetic damage of
Feingold syndrome, genetic damage of
Feingold Trainer syndrome, genetic damage of
Felty's syndrome, genetic damage of

Female pseudohermaphrodism, genetic damage of
Female pseudohermaphrodism Genuardi type, genetic damage of
Femoral facial syndrome, genetic damage of
Femur bifid monodactylous ectrodactyly, genetic damage of
Femur fibula ulna syndrome, genetic damage of
Fenton Wilkinson Toselano syndrome, genetic damage of
Ferlini Ragno Calzolari syndrome, genetic damage of
Fernhoff Blackston Oakley syndrome, genetic damage of
Ferrocalcinosis cerebro vascular, genetic damage of
Fetal akinesia syndrome X linked, genetic damage of
Fetal and neonatal alloimmune thrombocytopenia, genetic damage of
Fetal brain disruption sequence, genetic damage of
Fetal edema, genetic damage of
Fetal left ventricular aneurysm, genetic damage of
FG syndrome, genetic damage of
FGDY, genetic damage of
Fiber type disproportion, congenital, genetic damage of
Fibrinogen deficiency, congenital, genetic damage of
Fibrochondrogenesis, genetic damage of
Fibrofolliculomas with trichodiscomas and acrochordons, genetic damage of
Fibrolipomatosis, genetic damage of
Fibromatosis, genetic damage of
Fibromatosis gingival hypertrichosis, genetic damage of
Fibromatosis gingival progressive deafness, genetic damage of
Fibromatosis multiple non ossifying, genetic damage of
Fibromuscular dysplasia, genetic damage of
Fibromuscular dysplasia of arteries, genetic damage of
Fibromyalgia, genetic damage of
Fibrosarcoma, genetic damage of
Fibrosing alveolitis, genetic damage of
Fibrosis, genetic damage of

Fibrous dysplasia, genetic damage of
Fibrous dysplasia of bone, genetic damage of
Fibrousdysplasia ossificans progressiva, genetic damage of
Fibula aplasia complex brachydactyly, genetic damage of
Fibular aplasia ectrodactyly, genetic damage of
Fibular hypoplasia femoral bowing oligodactyly, genetic damage of
Fibular hypoplasia scapulo pelvic dysplasia absent, genetic damage of
Filippi syndrome, genetic damage of
Fine Lubinsky syndrome, genetic damage of
Fingerprints absence syndactyly milia, genetic damage of
Fingers absence, genetic damage of
Finnish congenital nephrosis, genetic damage of
Finnish lethal neonatal metabolic syndrome, genetic damage of
Finnish type amyloidosis, genetic damage of
Finucane Kurtz Scott syndrome, genetic damage of
Fish malodor syndrome, genetic damage of
Fish-eye disease, genetic damage of
Fissured tongue, genetic damage of
Fistulous vegetative verrucous hydradenoma, genetic damage of
Fitz-Hugh-Curtis syndrome, genetic damage of
Fitzsimmons Walson Mellor syndrome, genetic damage of
Fitzsimmons-Guilbert syndrome, genetic damage of
Fitzsimmons-McLachlan-Gilbert syndrome, genetic damage of
Flat foot, genetic damage of
Floating-harbor syndrome, genetic damage of
Florid cystic endosalpingiosis of the uterus, genetic damage of
Flotch syndrome, genetic damage of
Flynn Aird syndrome, genetic damage of
Focal agyria pachygyria, genetic damage of
Focal alopecia congenital megalencephaly, genetic damage of
Focal dermal hypoplasia, genetic damage of
Focal dystonia, genetic damage of

Focal or multifocal malformations in neuronal migration, genetic
damage of
Foix Chavany Marie syndrome, genetic damage of
Follicular atrophoderma-basal cell carcinoma, genetic damage of
Follicular hamartoma alopecia cystic fibrosis, genetic damage of
Follicular ichthyosis, genetic damage of
Follicular lymphoma, genetic damage of
Follicular lymphoreticuloma, genetic damage of
Fontaine Farriaux Blanckaert syndrome, genetic damage of
Forbes Albright syndrome, genetic damage of
Forbes disease, genetic damage of
Forestier's disease, genetic damage of
Forney Robinson Pascoe syndrome, genetic damage of
Fountain syndrome, genetic damage of
Fourth phacomatosis, genetic damage of
Fowler Christmas Chapele syndrome, genetic damage of
Fox-Fordyce disease, genetic damage of
Fragile X syndrome, genetic damage of
Fragile X syndrome type 1, genetic damage of
Fragile X syndrome type 2, genetic damage of
Fragile X syndrome type 3, genetic damage of
Fragoso Cid Garcia Hernandez syndrome, genetic damage of
Franceschetti-Klein syndrome, genetic damage of
Francheschini Vardeu Guala syndrome, genetic damage of
Francois dyscephalic syndrome, genetic damage of
Franek Bocker kahlen syndrome, genetic damage of
Fraser Jequier Chen syndrome, genetic damage of
Fraser like syndrome, genetic damage of
Fraser syndrome, genetic damage of
Frasier syndrome, genetic damage of
FRAXA syndrome, genetic damage of
FRAXD, genetic damage of
FRAXE syndrome, genetic damage of

Free sialic acid storage disease, genetic damage of
Freeman-Sheldon syndrome, genetic damage of
Freiberg's disease, genetic damage of
Freire Maia odontotrichomelic syndrome, genetic damage of
Freire Maia Pinheiro Opitz syndrome, genetic damage of
Frenkel Russe syndrome, genetic damage of
Frey's syndrome, genetic damage of
Frias syndrome, genetic damage of
Fried Goldberg Mundel syndrome, genetic damage of
Friedel Heid Grosshans syndrome, genetic damage of
Friedman Goodman syndrome, genetic damage of
Friedreich ataxia congenital glaucoma, genetic damage of
Friedreich's ataxia, genetic damage of
Frigophobia, genetic damage of (possible)
Froelich's syndrome, genetic damage of
Frölich's syndrome, genetic damage of
Fronto nasal malformation cloacal exstrophy, genetic damage of
Fronto-facio-nasal dysplasia, genetic damage of
Frontofacionasal dysplasia type Al gazali, genetic damage of
Frontometaphyseal dysplasia, genetic damage of
Frontonasal dysplasia, genetic damage of
Frontonasal dysplasia acromelic, genetic damage of
Frontonasal dysplasia klippel feil syndrome, genetic damage of
Frontonasal dysplasia phocomelic upper limbs, genetic damage of
Frontotemporal lobe dementia, genetic damage of
Froster huch syndrome, genetic damage of
Froster Iskenius Waterson syndrome, genetic damage of
Fructose intolerance, genetic damage of
Fructose-1,6-bisphosphatase deficiency, genetic damage of
Fructose-1-phosphate aldolase deficiency, hereditary, genetic damage of
Fructosemia, hereditary, genetic damage of
Fructosuria, genetic damage of

Frydman Cohen Ashenazi syndrome, genetic damage of
Frydman Cohen Karmon syndrome, genetic damage of
Fryer syndrome, genetic damage of
Fryns Fabry Remans syndrome, genetic damage of
Fryns Hofkens Fabry syndrome, genetic damage of
Fryns Smeets Thiry syndrome, genetic damage of
Fryns syndrome, genetic damage of
Fucosidosis, genetic damage of
Fucosidosis type 1, genetic damage of
Fuhrmann Rieger De sousa syndrome, genetic damage of
Fukuda Miyanomae Nakata syndrome, genetic damage of
Fukuyama type muscular dystrophy, genetic damage of
Fumarase deficiency, genetic damage of
Fumaric aciduria, genetic damage of
Fumarylacetoacetase deficiency, genetic damage of
Functioning pancreatic endocrine tumor, genetic damage of
Fuqua Berkovitz syndrome, genetic damage of
Furlong Kurczynski Hennessy syndrome, genetic damage of
Furukawa Takagi Nakao syndrome, genetic damage of
Fused mandibular incisors, genetic damage of

G

G syndrome, genetic damage of
Galactocerebrosidase deficiency, genetic damage of
Galactokinase deficiency, genetic damage of
Galactosamine-6-sulfatase deficiency, genetic damage of
Galactose-1-phosphate uridyltransferase deficiency, genetic damage of
Galactosemia, genetic damage of
Galactosialidosis, genetic damage of
Galloway Mowat syndrome, genetic damage of
Gamborg Nielsen syndrome, genetic damage of
Game Friedman Paradice syndrome, genetic damage of
Gamma aminobutyric acid transaminase deficiency, genetic damage of
Gamma-cystathionase deficiency, genetic damage of
Gamma-sarcoglycanopathy, genetic damage of
Gamophobia, genetic damage of (possible)
Gamstorp episodic adynamy, genetic damage of
Ganglioglioma, genetic damage of
Gangliosidosis (Type2)(GM2), genetic damage of
Gangliosidosis GM1 type 3, genetic damage of
Gangliosidosis type1, genetic damage of
Gapo syndrome, genetic damage of
Garcia Torres Guarner syndrome, genetic damage of
Gardner Morrisson Abbot syndrome, genetic damage of
Gardner Silengo Wachtel syndrome, genetic damage of
Gardner's syndrome, genetic damage of
Gardner-Diamond syndrome, genetic damage of
Garret Tripp syndrome, genetic damage of
Gastric lymphoma, genetic damage of
Gastritis, familial giant hypertrophic, genetic damage of
Gastro-enteropancreatic neuroendocrine tumor, genetic damage of
Gastrocutaneous syndrome, genetic damage of

Gastroenteritis, eosinophilic, genetic damage of
Gastroesophageal reflux, genetic damage of
Gastrointestinal autonomic nerve tumor, genetic damage of
Gastrointestinal neoplasm, genetic damage of
Gaucher Disease, genetic damage of
Gaucher disease type 1, genetic damage of
Gaucher disease type 2, genetic damage of
Gaucher disease type 3, genetic damage of
Gaucher ichthyosis restrictive dermopathy, genetic damage of
Gaucher-like disease, genetic damage of
Gay Feinmesser Cohen syndrome, genetic damage of
Geen Sandford Davison syndrome, genetic damage of
Gelatinous ascites, genetic damage of
Geleophysic dwarfism, genetic damage of
Gelineau disease, genetic damage of
Geliphobia, genetic damage of (possible)
Gemignani syndrome, genetic damage of
Gemss syndrome, genetic damage of
Generalized malformations in neuronal migration, genetic damage
of
Generalized resistance to thyroid hormone, genetic damage of
Generalized seizure, genetic damage of
Generalized torsion dystonia, genetic damage of
Genes syndrome, genetic damage of
Genetic reflex epilepsy, genetic damage of
Geniophobia, genetic damage of (possible)
Genital anomaly cardiomyopathy, genetic damage of
Genital dwarfism, genetic damage of
Genital dwarfism turner type, genetic damage of
Genito palatocardiac syndrome, genetic damage of
Genuphobia, genetic damage of (possible)
Geographic tongue, genetic damage of
Gerascophobia, genetic damage of (possible)

Gerhardt syndrome, genetic damage of
German syndrome, genetic damage of
Gerodermia osteodysplastica, genetic damage of
Gershinibaruch Leibo syndrome, genetic damage of
Gerstmann syndrome, genetic damage of
Gestational diabetes mellitus, genetic damage of
Gestational pemphigoid, genetic damage of
Gestational trophoblastic disease, genetic damage of
Ghosal syndrome, genetic damage of
Ghose Sachdev Kumar syndrome, genetic damage of
Gianotti-Crosti syndrome, genetic damage of
Giant axonal neuropathy, genetic damage of
Giant cell arteritis, genetic damage of
Giant cell myocarditis, genetic damage of
Giant congenital nevi, genetic damage of
Giant ganglionic hyperplasia, genetic damage of
Giant hypertrophic gastritis, genetic damage of
Giant mammary hamartoma, genetic damage of
Giant pigmented hairy nevus, genetic damage of
Giant platelet syndrome, genetic damage of
Gigantism, genetic damage of
Gigantism advanced bone age hoarse cry, genetic damage of
Gigantism exomphalos macroglossia, genetic damage of
Gigantism partial, nevi, hemihypertrophy, macrocephaly, genetic damage of
Gilbert's syndrome, genetic damage of
Gilles de la Tourette's syndrome, genetic damage of
Gingival fibromatosis dominant, genetic damage of
Gingival fibromatosis facial dysmorphism, genetic damage of
Gingival fibrosis, genetic damage of
Gingivitis, genetic damage of
Girate atrophy of choroid and retina, genetic damage of
Glanzmann thrombasthenia, genetic damage of

Glass Chapman Hockley syndrome, genetic damage of
Glastre Cochat Bouvier syndrome, genetic damage of
Glaucoma ecopia microspherophakia stiff joints short stature, genetic damage of
Glaucoma iridogoniodysgenesia, genetic damage of
Glaucoma sleep apnea, genetic damage of
Glaucoma type 1C, genetic damage of
Glaucoma, congenital, genetic damage of
Glaucoma, hereditary, genetic damage of
Glaucoma, hereditary adult type 1A, genetic damage of
Glaucoma, hereditary juvenile type 1B, genetic damage of
Glaucoma, primary infantile type 3A, genetic damage of
Glaucoma, primary infantile type 3B, genetic damage of
Glioblastoma, genetic damage of
Glioblastoma multiforme, genetic damage of
Glioma, genetic damage of
Gliomatosis cerebri, genetic damage of
Gliosarcoma, genetic damage of
Glomerulonephritis sparse hair telangiectases, genetic damage of
Glomerulosclerosis, genetic damage of
Gloomy face syndrome, genetic damage of
Glossodynia, genetic damage of
Glossopalatine ankylosis cataracts digital anomalies, genetic damage of
Glossopalatine ankylosis micrognathia ear anomalies, genetic damage of
Glossopharyngeal neuralgia, genetic damage of
Glossophobia, genetic damage of (possible)
Glucagonoma, genetic damage of
Glucocorticoid deficiency, familial, genetic damage of
Glucocorticoid resistance, genetic damage of
Glucocorticoid sensitive hypertension, genetic damage of
Glucose 6 phosphate dehydrogenase deficiency, genetic damage of

Glucose-6-phosphate translocase deficiency, genetic damage of
Glucose-galactose malabsorption, genetic damage of
Glucosephosphate isomerase deficiency, genetic damage of
Glucosidase acid-1,4-alpha deficiency, genetic damage of
Glut2 deficiency, genetic damage of
Glutamate decarboxylase deficiency, genetic damage of
Glutamate-aspartate transport defect, genetic damage of
Glutaricaciduria I, genetic damage of
Glutaricaciduria II, genetic damage of
Glutaryl-CoA dehydrogenase deficiency, genetic damage of
Glyceraldehyde-3-phosphate dehydrogenase deficiency, genetic damage of
Glycerol kinase deficiency, genetic damage of
Glycine synthase deficiency, genetic damage of
Glycinemia, ketotic, genetic damage of
Glycogen storage disease type 1A, genetic damage of
Glycogen storage disease type 1B, genetic damage of
Glycogen storage disease type 1C, genetic damage of
Glycogen storage disease type 1D, genetic damage of
Glycogen storage disease type 6, due to phosphorylation, genetic damage of
Glycogen storage disease type 7, genetic damage of
Glycogen storage disease type 9, genetic damage of
Glycogen storage disease type I, genetic damage of
Glycogen storage disease type II, genetic damage of
Glycogen storage disease type V, genetic damage of
Glycogen storage disease type VI, genetic damage of
Glycogen storage disease type VII, genetic damage of
Glycogen storage disease type VIII, genetic damage of
Glycogenosis type II, genetic damage of
Glycogenosis type III, genetic damage of
Glycogenosis type IV, genetic damage of
Glycogenosis type V, genetic damage of

Glycogenosis type VI, genetic damage of
Glycogenosis type VII, genetic damage of
Glycogenosis type VIII, genetic damage of
Glycogenosis, type 0, genetic damage of
Glycosuria, genetic damage of
GM2 gangliosidosis, 0 variant, genetic damage of
GM2-gangliosidosis, B, B1, AB variant, genetic damage of
GMS syndrome, genetic damage of
Goiter-deafness syndrome, genetic damage of
Goldberg Bull syndrome, genetic damage of
Goldberg syndrome, genetic damage of
Goldblatt Wallis syndrome, genetic damage of
Goldblatt Wallis Zieff syndrome, genetic damage of
Goldblatt Viljoen syndrome, genetic damage of
Goldenhar disease, genetic damage of
Goldskag Cooks Hertz syndrome, genetic damage of
Goldstein Hutt syndrome, genetic damage of
Gollop Coates syndrome, genetic damage of
Gollop syndrome, genetic damage of
Gollop Wolfgang complex, genetic damage of
Goltz syndrome, genetic damage of
Gombo syndrome, genetic damage of
Gomez and López-Hernández syndrome, genetic damage of
Gonadal dysgenesis, genetic damage of
Gonadal dysgenesis mixed, genetic damage of
Gonadal dysgenesis Turner type, genetic damage of
Gonadal dysgenesis XY type associated anomalies, genetic damage of
Gonadal dysgenesis, XX type, genetic damage of
Gonadal dysgenesis, XY female type, genetic damage of
Goniodysgenesis mental retardation short stature, genetic damage of
Gonococcal conjunctivitis, genetic damage of

Gonodal dysgenesis XX type deafness, genetic damage of
Gonzales Del Angel syndrome, genetic damage of
Goodman camptodactyly, genetic damage of
Goodman syndrome, genetic damage of
Goodpasture pneumorenal syndrome, genetic damage of
Goodpasture's syndrome, genetic damage of
Gordon hyperkaliemia-hypertension syndrome, genetic damage of
Gordon syndrome, genetic damage of
Gorham syndrome, genetic damage of
Gorham-Stout disease, genetic damage of
Gorlin Bushkell Jensen syndrome, genetic damage of
Gorlin Chaudhry Moss syndrome, genetic damage of
Gottron's syndrome, genetic damage of
Gougerot-Sjogren syndrome, genetic damage of
Gouty nephropathy, familial, genetic damage of
Graft versus host disease, genetic damage of
Graham Boyle Troxell syndrome, genetic damage of
Grand Kaine Fulling syndrome, genetic damage of
Grant syndrome, genetic damage of
Granulocytopenia, genetic damage of
Granuloma annulare, genetic damage of
Granulomatosis, lymphomatoid, genetic damage of
Granulomatous allergic angiitis, genetic damage of
Granulomatous hypophysitis, genetic damage of
Granulomatous rosacea, genetic damage of
Graves' disease, genetic damage of
Gray platelet syndrome, genetic damage of
Great vessels transposition, genetic damage of
Greenberg dysplasia, genetic damage of
Greig cephalopolysyndactyly syndrome, genetic damage of
Greig syndrome, genetic damage of
Griscelli disease, genetic damage of
Grix Blankenship Peterson syndrome, genetic damage of

Groll Hirschowitz syndrome, genetic damage of
Grosse syndrome, genetic damage of
Grover's disease, genetic damage of
Growth deficiency brachydactyly unusual facies, genetic damage of
Growth delay, constitutional, genetic damage of
Growth hormone deficiency, genetic damage of
Growth mental deficiency syndrome of Myhre, genetic damage of
Growth retardation alopecia pseudoanodontia optic, genetic damage of
Growth retardation hydrocephaly lung hypoplasia, genetic damage of
Growth retardation mental retardation phalangeal hypoplasia, genetic damage of
Grubben Decock Borghgraef syndrome, genetic damage of
GTP cyclohydrolase deficiency, genetic damage of
Guanidinoacetate methyltransferase deficiency, genetic damage of
Guibaud Vainsel syndrome, genetic damage of
Guillan-Barre syndrome, genetic damage of
Guizar Vasquez Luengas syndrome, genetic damage of
Guizar Vasquez Sanchez Manzano syndrome, genetic damage of
Gunal Seber Basaran syndrome, genetic damage of
Gupta Patton syndrome, genetic damage of
Gurrieri Sammito Bellussi syndrome, genetic damage of
Gusher syndrome, genetic damage of
Gymnophobia, genetic damage of (possible)
Gyrate atrophy, genetic damage of
Gyrate atrophy of the retina, genetic damage of

H

Haas Chir Robinson syndrome, genetic damage of
Haemangioendothelioma, genetic damage of
Haemorrhagic proctocolitis, genetic damage of
Hageman factor deficiency, genetic damage of
Hagemoser Weinstein Bresnick syndrome, genetic damage of
Hagiophobia, genetic damage of (possible)
Hailey-Hailey disease, genetic damage of
Hair defect photosensitivity mental retardation, genetic damage of
Hairy cell leukemia, genetic damage of
Hairy throat syndrome, genetic damage of
Hairy tongue, genetic damage of
Hajdu-Cheney syndrome, genetic damage of
Halal Setton Wang syndrome, genetic damage of
Halal syndrome, genetic damage of
Hall Riggs mental retardation syndrome, genetic damage of
Hallermann Streiff Francois syndrome, genetic damage of
Hallermann-Streiff syndrome, genetic damage of
Hallervorden-Spatz disease, genetic damage of
Hallux varus and preaxial polysyndactyly, genetic damage of
Hamanishi Ueba Tsuji syndrome, genetic damage of
Hamano Tsukamoto syndrome, genetic damage of
Hamartoma sebaceus of Jadassohn, genetic damage of
Hamman-Rich disease, genetic damage of
Hand and foot deformity flat facies, genetic damage of
Hand foot uterus syndrome, genetic damage of
Hand wringing Rett syndrome, genetic damage of
Hand-foot-mouth disease, genetic damage of
Hand-Schuller-Christian disease, genetic damage of
Hanhart syndrome, genetic damage of
Hapnes Boman Skeie syndrome, genetic damage of
Hard skin syndrome Parana type, genetic damage of
HARD syndrome, genetic damage of

Harding ataxia, genetic damage of
Harlequin type ichthyosis, genetic damage of
Harpaxophobia, genetic damage of (possible)
Harrod Doman Keele syndrome, genetic damage of
Hartnup disease, genetic damage of
Hartsfield Bixler Demyer syndrome, genetic damage of
Hashimoto struma, genetic damage of
Hashimoto's syndrome, genetic damage of
Hashimoto-Pritzker syndrome, genetic damage of
Haspeslagh Fryns Muelenaere syndrome, genetic damage of
Hay Wells syndrome recessive type, genetic damage of
Hay-Wells syndrome, genetic damage of
Headache, cluster, genetic damage of
Hearing disorder, genetic, genetic damage of
Hearing impairment, genetic, genetic damage of
Hearing loss, genetic, genetic damage of
Heart aneurysm, genetic damage of
Heart block, genetic damage of
Heart block progressive, familial, genetic damage of
Heart defect round face congenital retarded development, genetic damage of
Heart defect tongue hamartoma polysyndactyly, genetic damage of
Heart defects limb shortening, genetic damage of
Heart hand syndrome Spanish type, genetic damage of
Heart hypertrophy, hereditary, genetic damage of
Heart situs anomaly, genetic damage of
Heart tumor of the adult, genetic damage of
Heart tumor of the child, genetic damage of
HEC syndrome, genetic damage of
Hecht Scott syndrome, genetic damage of
Heckenlively syndrome, genetic damage of
Heide syndrome, genetic damage of
Heliophobia, genetic damage of (possible)

Helmerhorst Heaton Crossen syndrome, genetic damage of
HEM dysplasia, genetic damage of
Hemangioblastoma, genetic damage of
Hemangioma, genetic damage of
Hemangioma thrombocytopenia syndrome, genetic damage of
Hemangiomas cavernous of face supraumbilical midline raphe, genetic damage of
Hemangiopericytoma, genetic damage of
Hemeralopia, congenital essential, genetic damage of
Hemeralopia, familial, genetic damage of
Hemi 3 syndrome, genetic damage of
Hemifacial atrophy agenesis of the caudate nucleus, genetic damage of
Hemifacial atrophy progressive, genetic damage of
Hemifacial hyperplasia strabismus, genetic damage of
Hemifacial microsomia, genetic damage of
Hemihypertrophy in context of NF, genetic damage of
Hemihypertrophy intestinal web corneal opacity, genetic damage of
Hemimegalencephaly, genetic damage of
Hemiplegia, genetic damage of
Hemiplegic migraine, familial, genetic damage of
Hemoglobin C disease, genetic damage of
Hemoglobin E disease, genetic damage of
Hemoglobin SC disease, genetic damage of
Hemoglobinopathy, genetic damage of
Hemoglobinuria, genetic damage of
Hemolytic anemia lethal genital anomalies, genetic damage of
Hemolytic-uremic syndrome, genetic damage of
Hemophagocytic lymphohistiocytosis, genetic damage of
Hemophagocytic reticulosis (or reticulitis), genetic damage of
Hemophilia, genetic damage of
Hemophilic arthropathy, genetic damage of

Hemophobia, genetic damage of (possible)
Hemorrhagic thrombocythemia, genetic damage of
Hemorrhagiparous thrombocytic dystrophy, genetic damage of
Hemosiderosis, genetic damage of
Hennekam Beemer syndrome, genetic damage of
Hennekam Koss de Geest syndrome, genetic damage of
Hennekam syndrome, genetic damage of
Hennekam Van der Horst syndrome, genetic damage of
Heparane sulfamidase deficiency, genetic damage of
Hepatic cystic hamartoma, genetic damage of
Hepatic ductular hypoplasia, genetic damage of
Hepatic fibrosis, genetic damage of
Hepatic fibrosis renal cysts mental retardation, genetic damage of
Hepatic venoocclusive disease, genetic damage of
Hepatoblastoma, genetic damage of
Hepatocellular carcinoma, genetic damage of
Hepatorenal syndrome, genetic damage of
Hepatorenal tyrosinemia, genetic damage of
Hereditary amyloidosis, genetic damage of
Hereditary angioedema, genetic damage of
Hereditary ataxia, genetic damage of
Hereditary carnitine deficiency, genetic damage of
Hereditary carnitine deficiency myopathy, genetic damage of
Hereditary ceroid-lipofuscinosis, genetic damage of
Hereditary coproporphyria, genetic damage of
Hereditary deafness, genetic damage of
Hereditary elliptocytosis, genetic damage of
Hereditary fructose intolerance, genetic damage of
Hereditary hearing disorder, genetic damage of
Hereditary hearing loss, genetic damage of
Hereditary hemochromatosis, genetic damage of
Hereditary hemorrhagic telangiectasia, genetic damage of
Hereditary hyperuricemia, genetic damage of

Hereditary lymphedema, genetic damage of
Hereditary macrothrombocytopenia, genetic damage of
Hereditary methemoglobinemia, recessive, genetic damage of
Hereditary motor and sensory neuropathy, genetic damage of
Hereditary myopathy with intranuclear filamentous, genetic damage of
Hereditary nodular heterotopia, genetic damage of
Hereditary non-spherocytic hemolytic anemia, genetic damage of
Hereditary pancreatitis, genetic damage of
Hereditary paroxysmal cerebral ataxia, genetic damage of
Hereditary peripheral nervous disorder, genetic damage of
Hereditary primary Fanconi disease, genetic damage of
Hereditary resistance to anti-vitamin K, genetic damage of
Hereditary sensory and autonomic neuropathy 3, genetic damage of
Hereditary sensory and autonomic neuropathy 4, genetic damage of
Hereditary sensory neuropathy type I, genetic damage of
Hereditary sensory neuropathy type II, genetic damage of
Hereditary spastic paraplegia, genetic damage of
Hereditary spherocytic hemolytic anemia, genetic damage of
Hereditary spherocytosis, genetic damage of
Hereditary type 1 neuropathy, genetic damage of
Hereditary type 2 neuropathy, genetic damage of
Hereditary tyrosinemia, genetic damage of
Hermansky-Pudlak syndrome, genetic damage of
Hermaphroditism, genetic damage of
Hernandez Aguire Negrete syndrome, genetic damage of
Herpetophobia, genetic damage of (possible)
Herrmann Opitz arthrogryposis syndrome, genetic damage of
Herrmann Opitz craniosynostosis, genetic damage of
Hers disease, genetic damage of
Hersh Podruch Weisskopk syndrome, genetic damage of

Heterophobia, genetic damage of (possible)
Heterotaxia (generic term), genetic damage of
Heterotaxia autosomal dominant type, genetic damage of
Heterotaxy with polysplenia or asplenia, genetic damage of
Heterotaxy, visceral, X linked, genetic damage of
Hexosaminidase A deficiency, genetic damage of
Hexosaminidases A and B deficiency, genetic damage of
HHH syndrome, genetic damage of
Hidradenitis suppurativa, genetic damage of
Hidradenitis suppurativa familial, genetic damage of
Hidrotic ectodermal dysplasia type Christianson Fouris, genetic damage of
High scapula, genetic damage of
High-molecular-weight kininogen deficiency, congenital, genetic damage of
Hillig syndrome, genetic damage of
Hing Torack Dowston syndrome, genetic damage of
Hinson-Pepys disease, genetic damage of
Hip dislocation, genetic damage of
Hip Dysplasia, genetic damage of
Hip dysplasia Beukes type, genetic damage of
Hip luxation, genetic damage of
Hip subluxation, genetic damage of
Hipo syndrome, genetic damage of
Hippel Lindau disease, genetic damage of
Hirschsprung disease ganglioneuroblastoma, genetic damage of
Hirschsprung disease polydactyly heart disease, genetic damage of
Hirschsprung disease type 2, genetic damage of
Hirschsprung disease type 3, genetic damage of
Hirschsprung disease type d brachydactyly, genetic damage of
Hirschsprung disease with pigmentary anomaly, genetic damage of
Hirschsprung microcephaly cleft palate, genetic damage of
Hirschsprung nail hypoplasia dysmorphism, genetic damage of

Hirschsprung's disease, genetic damage of
Hirsutism congenital gingival hyperplasia, genetic damage of
Hirsutism skeletal dysplasia mental retardation, genetic damage of
His bundle tachycardia, genetic damage of
Histidase deficiency, genetic damage of
Histidinemia, genetic damage of
Histidinuria renal tubular defect, genetic damage of
Histiocytosis X, genetic damage of
Histiocytosis, Non-Langerhans-Cell, genetic damage of
Hittner Hirsch Kreh syndrome, genetic damage of
Hm syndrome, genetic damage of
HMC syndrome, genetic damage of
Hmg coa synthetase deficiency, genetic damage of
Hodgkin lymphoma, genetic damage of
Hodgkin's disease, genetic damage of
Hodophobia, genetic damage of (possible)
Hoepffner Dreyer Reimers syndrome, genetic damage of
Hollow visceral myopathy, genetic damage of
Holmes Benacerraf syndrome, genetic damage of
Holmes Borden syndrome, genetic damage of
Holmes Collins syndrome, genetic damage of
Holmes Gang syndrome, genetic damage of
Holoacardius amorphus, genetic damage of
Holocarboxylase synthetase deficiency, genetic damage of
Holoprosencephaly, genetic damage of
Holoprosencephaly caudal dysgenesis, genetic damage of
Holoprosencephaly craniosynostosis, genetic damage of
Holoprosencephaly deletion 2p, genetic damage of
Holoprosencephaly ectrodactyly cleft lip palate, genetic damage of
Holoprosencephaly radial heart renal anomalies, genetic damage of
Holt-Oram syndrome, genetic damage of
Holzgreve Wagner Rehder syndrome, genetic damage of
Homocarnosinase deficiency, genetic damage of

Homocarnosinosis, genetic damage of

Homocystinuria, genetic damage of

Homocystinuria due to cystathionine beta-synthase, genetic damage of

Homocystinuria due to defect in methylation (cbl g), genetic damage of

Homocystinuria due to defect in methylation cbl e, genetic damage of

Homocystinuria due to defect in methylation, MTHFR deficiency, genetic damage of

Homozygous hypobetalipoproteinemia, genetic damage of

Hoon Hall syndrome, genetic damage of

Hordnes Engebretsen Knudtson syndrome, genetic damage of

Horn Kolb syndrome, genetic damage of

Horner's syndrome, genetic damage of

Hornova Dlurosova syndrome, genetic damage of

Horseshoe kidney, genetic damage of

Horton disease, genetic damage of

Horton disease, juvenile, genetic damage of

Houlston Ironton Temple syndrome, genetic damage of

Howard Young syndrome, genetic damage of

Howell-Evans syndrome, genetic damage of

Hoyeraal Hreidarsson syndrome, genetic damage of

Hoyeraal syndrome, genetic damage of

Humero spinal dysostosis congenital heart disease, genetic damage of

Humeroradial synostosis, genetic damage of

Humeroradioulnar synostosis, genetic damage of

Humerus trochlea, aplasia of, genetic damage of

Hunter Carpenter McDonald syndrome, genetic damage of

Hunter Jurenka Thompson syndrome, genetic damage of

Hunter Macpherson syndrome, genetic damage of

Hunter McAlpine syndrome, genetic damage of

Hunter McDonald syndrome, genetic damage of
Hunter Rudd Hoffmann syndrome, genetic damage of
Hunter syndrome, genetic damage of
Hunter Thomson Reed syndrome, genetic damage of
Huntington's disease, genetic damage of
Huriez scleroatrophic syndrome Hurler syndrome, genetic damage of
Hurler syndrome, genetic damage of
Hurst Hallam Hockey syndrome, genetic damage of
Hutchinson Gilford Progeria syndrome, genetic damage of
Hutchinson incisors, genetic damage of
Hutchinson-Gilford syndrome, genetic damage of
Hutteroth Spranger syndrome, genetic damage of
Hyalinosis systemic short stature, genetic damage of
Hyde Forster McCarthy Berry syndrome, genetic damage of
Hydranencephaly, genetic damage of
Hydrocephalus, genetic damage of
Hydrocephalus—Arnold Chiari—allied disorders, genetic damage of
Hydrocephalus autosomal recessive, genetic damage of
Hydrocephalus cataract microphthalmos, genetic damage of
Hydrocephalus costovertebral dysplasia Sprengel anomaly, genetic damage of
Hydrocephalus craniosynostosis bifid nose, genetic damage of
Hydrocephalus endocardial fibroelastosis cataract, genetic damage of
Hydrocephalus growth retardation skeletal anomalies, genetic damage of
Hydrocephalus obesity hypogonadism, genetic damage of
Hydrocephalus skeletal anomalies, genetic damage of
Hydrocephaly corpus callosum agenesis diaphragmatic hernia, genetic damage of
Hydrocephaly low insertion umbilicus, genetic damage of

Hydrocephaly tall stature joint laxity, genetic damage of
Hydrolethalus syndrome, genetic damage of
Hydronephrosis congenital, genetic damage of
Hydronephrosis peculiar facial expression, genetic damage of
Hydrophobia, genetic damage of (possible)
Hydrops ectrodactyly syndactyly, genetic damage of
Hydrops fetalis, genetic damage of
Hydrops fetalis anemia immune disorder absent thumb, genetic damage of
Hydroxymethylglutaricaciduria, genetic damage of
Hygroma cervical, genetic damage of
Hyper IgM syndrome, genetic damage of
Hyper-IgD syndrome, genetic damage of
Hyper-reninism, genetic damage of
Hyperadrenalism, genetic damage of
Hyperaldosteronism, genetic damage of
Hyperaldosteronism familial type 2, genetic damage of
Hyperaldosteronism, familial type 1, genetic damage of
Hyperammonemia, genetic damage of
Hyperandrogenism, genetic damage of
Hyperbilirubinemia, genetic damage of
Hyperbilirubinemia transient familial neonatal, genetic damage of
Hyperbilirubinemia type 1, genetic damage of
Hyperbilirubinemia type 2, genetic damage of
Hypercalcemia, genetic damage of
Hypercalcemia, familial benign, genetic damage of
Hypercalcemia, familial benign type 1, genetic damage of
Hypercalcemia, familial benign type 2, genetic damage of
Hypercalcemia, familial benign type 3, genetic damage of
Hypercalciuria, genetic damage of
Hypercalciuria idiopathic, genetic damage of
Hypercalciuria macular coloboma, genetic damage of
Hypercementosis, genetic damage of

Hypercholesterolemia, genetic damage of

Hypercholesterolemia due to arg3500 mutation of Apo B-100, genetic damage of

Hypercholesterolemia due to LDL receptor deficiency, genetic damage of

Hyperchylomicronemia, genetic damage of

Hypereosinophilic syndrome, genetic damage of

Hyperferritinemia, hereditary, with congenital cataracts, genetic damage of

Hyperglycerolemia, genetic damage of

Hyperglycinemia, genetic damage of

Hyperglycinemia, isolated nonketotic, genetic damage of

Hyperglycinemia, isolated nonketotic type 1, genetic damage of

Hyperglycinemia, isolated nonketotic type 2, genetic damage of

Hypergonadotropic ovarian failure, familial or sporadic, genetic damage of

Hyperhidrosis, genetic damage of

Hyperhomocysteinemia, genetic damage of

Hyperimidodipeptiduria, genetic damage of

Hyperimmunoglobinemia D with recurrent fever, genetic damage of

Hyperimmunoglobulin E—reccurrent infection syndrome, genetic damage of

Hyperimmunoglobulinemia D with periodic fever, genetic damage of

Hyperimmunoglobulinemia E, genetic damage of

Hyperinsulinism due to focal adenomatous hyperplasia, genetic damage of

Hyperinsulinism due to glucokinase deficiency, genetic damage of

Hyperinsulinism due to glutamodehydrogenase deficiency, genetic damage of

Hyperinsulinism in children, congenital, genetic damage of

Hyperinsulinism, diffuse, genetic damage of

Hyperinsulinism, focal, genetic damage of
Hyperkalemia, genetic damage of
Hyperkalemic periodic paralysis, genetic damage of
Hyperkeratosis lenticularis perstans, genetic damage of
Hyperkeratosis lenticularis perstans of Flegel, genetic damage of
Hyperkeratosis palmoplantar localized acanthokeratolytic, genetic
damage of
Hyperkeratosis palmoplantar localized epidermolytic, genetic
damage of
Hyperkeratosis palmoplantar with palmar crease hyperkeratosis,
genetic damage of
Hyperlipoproteinemia, genetic damage of
Hyperlipoproteinemia type I, genetic damage of
Hyperlipoproteinemia type II, genetic damage of
Hyperlipoproteinemia type III, genetic damage of
Hyperlipoproteinemia type V, genetic damage of
Hyperlysinemia, genetic damage of
Hyperornithinemia, genetic damage of
Hyperornithinemia-hyperammonemia-homocitrullinuria, genetic
damage of
Hyperostosid corticalis deformans juvenilis, genetic damage of
Hyperostosis cortical infantile, genetic damage of
Hyperostosis corticalis generalisata, genetic damage of
Hyperostosis frontalis interna, genetic damage of
Hyperostosis generalisata with striations, genetic damage of
Hyperoxaluria, genetic damage of
Hyperoxaluria type 1, genetic damage of
Hyperoxaluria type 2, genetic damage of
Hyperparathyroidism, genetic damage of
Hyperparathyroidism, familial, primary, genetic damage of
Hyperparathyroidism, neonatal severe primary, genetic damage of
Hyperphalangism dysmorphy bronchomalacia, genetic damage of
Hyperphenilalaninemia due to pterin-4-alpha-carbin, genetic

damage of

Hyperphenylalalinemia due to dihydropteridine reductase deficiency, genetic damage of

Hyperphenylalaninemia due to 6-pyruvoyltetrahydrop, genetic damage of

Hyperphenylalaninemia due to dehydratase deficiency, genetic damage of

Hyperphenylalaninemia due to GTP cyclohydrolase deficiency, genetic damage of

Hyperphenylalaninemic embryopathy, genetic damage of

Hyperpipecolatemia, genetic damage of

Hyperprolactinemia, genetic damage of

Hyperprolinemia, genetic damage of

Hyperprolinemia type I, genetic damage of

Hyperprolinemia type II, genetic damage of

Hyperreflexia, genetic damage of

Hypersomnolence, genetic damage of

Hypertelorism and tetralogy of Fallot, genetic damage of

Hypertelorism hypospadias syndrome, genetic damage of

Hypertelorism microtia facial clefting syndrome, genetic damage of

Hypertension, genetic damage of

Hypertensive hyperkalemia, familial, genetic damage of

Hypertensive hypokalemia familial, genetic damage of

Hypertensive retinopathy, genetic damage of

Hyperthermia, genetic damage of

Hyperthermia induced defects, genetic damage of

Hyperthermia of anesthesia, genetic damage of

Hyperthyroidism due to mutations in TSH receptor, genetic damage of

Hypertrichosis atrophic skin ectropion macrostomia, genetic damage of

Hypertrichosis brachydactyly obesity and mental retardation,

genetic damage of
Hypertrichosis congenital generalized X linked, genetic damage of
Hypertrichosis lanuginosa, genetic damage of
Hypertrichosis retinopathy dysmorphism, genetic damage of
Hypertrichosis universalis congenita Ambras type, genetic damage of
Hypertrichotic osteochondrodysplasia, genetic damage of
Hypertrophic branchial myopathy, genetic damage of
Hypertrophic cardiomyopathy, genetic damage of
Hypertrophic hemangiectasia, genetic damage of
Hypertrophic myocardiopathy, genetic damage of
Hypertrophic osteoarthropathy, primary or idiopathic, genetic damage of
Hypertropic neuropathy of Dejerine-Sottas, genetic damage of
Hypertryptophanemia, genetic damage of
Hypo-alphalipoproteinemia primary, genetic damage of
Hypoadrenalism, genetic damage of
Hypoadrenocorticism hypoparathyroidism moniliasis, genetic damage of
Hypoaldosteronism, genetic damage of, genetic damage of
Hypobetalipoproteinaemia ataxia hearing loss, genetic damage of
Hypobetalipoprotéinemia, familial, genetic damage of
Hypocalcemia, genetic damage of
Hypocalcemia, autosomal dominant, genetic damage of
Hypocalciuric hypercalcemia, familial, genetic damage of
Hypocalciuric hypercalcemia, familial type 1, genetic damage of
Hypocalciuric hypercalcemia, familial type 2, genetic damage of
Hypocalciuric hypercalcemia, familial type 3, genetic damage of
Hypochondrogenesis, genetic damage of
Hypochondroplasia, genetic damage of
Hypocomplementemic urticarial vasculitis, genetic damage of
Hypodermyasis, genetic damage of
Hypodontia dysplasia of nails, genetic damage of

Hypodontia of incisors and premolars, genetic damage of
Hypofibrinogenemia, familial, genetic damage of
Hypoglycemia with deficiency of glycogen synthetase in the liver, genetic damage of
Hypogonadism, genetic damage of
Hypogonadism cardiomyopathy, genetic damage of
Hypogonadism hypogonadotropic due to mutations in GR hormone, genetic damage of
Hypogonadism male mental retardation skeletal anomaly, genetic damage of
Hypogonadism mitral valve prolapse mental retardation, genetic damage of
Hypogonadism primary partial alopecia, genetic damage of
Hypogonadism retinitis pigmentosa, genetic damage of
Hypogonadism, isolated, hypogonadotropic, genetic damage of
Hypogonadotropic hypogonadism syndactyly, genetic damage of
Hypogonadotropic hypogonadism without anosmia, X linked, genetic damage of
Hypogonadotropic hypogonadism-anosmia, genetic damage of
Hypogonadotropic hypogonadism-anosmia, X linked, genetic damage of
Hypohidrotic Ectodermal Dysplasia, genetic damage of
Hypokalemia, genetic damage of
Hypokalemic alkalosis with hypercalciuria, genetic damage of
Hypokalemic periodic paralysis, genetic damage of
Hypokaliemic periodic paralysis type 1, genetic damage of
Hypoketonemic hypoglycemia, genetic damage of
Hypokinetic dilated cardiomyopathy, familial, genetic damage of
Hypomagnesemia primary, genetic damage of
Hypomandibular faciocranial dysostosis, genetic damage of
Hypomelanotic disorder, genetic damage of
Hypomelia mullerian duct anomalies, genetic damage of
Hypomentia, genetic damage of

Hypoparathyroidism, genetic damage of
Hypoparathyroidism familial isolated, genetic damage of
Hypoparathyroidism nerve deafness nephrosis, genetic damage of
Hypoparathyroidism short stature, genetic damage of
Hypoparathyroidism short stature mental retardation, genetic damage of
Hypoparathyroidism X linked, genetic damage of
Hypophosphatasia, genetic damage of
Hypophosphatasia, infantile, genetic damage of
Hypophosphatemic rickets, genetic damage of
Hypopigmentation oculocerebral syndrome Cross type, genetic damage of
Hypopituitarism, genetic damage of
Hypopituitarism micropenis cleft lip palate, genetic damage of
Hypopituitarism microphthalmia, genetic damage of
Hypopituitarism postaxial polydactyly, genetic damage of
Hypopituitary dwarfism, genetic damage of
Hypoplasia hepatic ductular, genetic damage of
Hypoplasia of the tibia with polydactyly, genetic damage of
Hypoplastic left heart syndrome, genetic damage of
Hypoplastic right heart microcephaly, genetic damage of
Hypoplastic thumb mullerian aplasia, genetic damage of
Hypoplastic thumbs hydranencephaly, genetic damage of
Hypoproconvertinemia, genetic damage of
Hypoprothrombinemia, genetic damage of
Hyporeninemic hypoaldosteronism, genetic damage of
Hyposmia nasal hypoplasia hypogonadism, genetic damage of
Hypospadias familial, genetic damage of
Hypospadias mental retardation Goldblatt type, genetic damage of
Hypotelorism cleft palate hypospadias, genetic damage of
Hypotension, orthostatic, genetic damage of (possible)
Hypothalamic dysfunction, genetic damage of
Hypothalamic hamartoblastoma syndrome, genetic damage of

Hypothalamic hamartomas, genetic damage of
Hypothyroidism, genetic damage of
Hypothyroidism due to iodide transport defect, genetic damage of
Hypothyroidism postaxial polydactyly mental retardation, genetic damage of
Hypotonic sclerotic muscular dystrophy, genetic damage of
Hypotrichosis, genetic damage of
Hypotrichosis mental retardation Lopes type, genetic damage of
Hypoxanthine guanine phosphoribosyltransferase deficiency, genetic damage of

I

I-cell disease, genetic damage of
IBIDS syndrome, genetic damage of
ICF syndrome, genetic damage of
Ichthyophobia, genetic damage of (possible)
Ichthyosiform erythroderma corneal involvement deafness, genetic damage of
Ichthyosis alopecia eclabion ectropion mental retardation, genetic damage of
Ichthyosis and male hypogonadism, genetic damage of
Ichthyosis bullosa of Siemens, genetic damage of
Ichthyosis cheek eyebrow syndrome, genetic damage of
Ichthyosis congenita, genetic damage of
Ichthyosis congenita biliary atresia, genetic damage of
Ichthyosis congenita, collodion fetus type, genetic damage of
Ichthyosis deafness mental retardation skeletal anomaly, genetic damage of
Ichthyosis exfoliativa, genetic damage of
Ichthyosis follicularis atrichia photophobia syndrome, genetic damage of
Ichthyosis hepatosplenomegaly cerebellar degeneration, genetic damage of
Ichthyosis hystrix, Curth Macklin type, genetic damage of
Ichthyosis linearis circumflexa, genetic damage of
Ichthyosis male hypogonadism, genetic damage of
Ichthyosis mental retardation Devriendt type, genetic damage of
Ichthyosis mental retardation dwarfism renal impairment, genetic damage of
Ichthyosis microphthalmos, genetic damage of
Ichthyosis tapered fingers midline groove up, genetic damage of
Ichthyosis vulgaris, genetic damage of
Ichthyosis, erythrokeratolysis hemalis, genetic damage of
Ichthyosis, keratosis follicularis spinulosa Decalvans, genetic dam-

age of
Ichthyosis, lamellar recessive, genetic damage of
Ichthyosis, Netherton syndrome, genetic damage of
Idaho syndrome, genetic damage of
Idiopathic adolescent scoliosis, genetic damage of
Idiopathic adult neutropenia, genetic damage of
Idiopathic alveolar hypoventilation syndrome, genetic damage of
Idiopathic congenital nystagmus, dominant, X linked, genetic
damage of
Idiopathic diffuse interstitial fibrosis, genetic damage of
Idiopathic dilatation of the pulmonary artery, genetic damage of
Idiopathic dilation cardiomyopathy, genetic damage of
Idiopathic double athetosis, genetic damage of
Idiopathic edema, genetic damage of
Idiopathic eosinophilic chronic pneumopathy, genetic damage of
Idiopathic facial palsy, genetic damage of
Idiopathic hypereosinophilic syndrome, genetic damage of
Idiopathic juvenile osteoporosis, genetic damage of
Idiopathic orthostatic hypotension, genetic damage of
Idiopathic pulmonary fibrosis, genetic damage of
Idiopathic pulmonary hemosiderosis, genetic damage of
Idiopathic sclerosing mesenteritis, genetic damage of
Idiopathic thrombocytopenic purpura, genetic damage of
Iduronate 2-sulfatase deficiency, genetic damage of
IFAP syndrome, genetic damage of
IgA deficiency, genetic damage of
IgA nephropathy, genetic damage of
IGDA syndrome, genetic damage of
Illum syndrome, genetic damage of
Illyngophobia, genetic damage of (possible)
Ilyina Amoashy Grygory syndrome, genetic damage of
Imaizumi Kuroki syndrome, genetic damage of
Immotile cilia syndrome, due to defective radial spokes, genetic

damage of

Immotile cilia syndrome, due to excessively long cilia, genetic damage of

Immotile cilia syndrome, Kartagener type, genetic damage of

Immune deficiency, familial variable, genetic damage of

Immune thrombocytopenia, genetic damage of

Immunodeficiency microcephaly and chromosomal instability, genetic damage of

Immunodeficiency with short limb dwarfism, genetic damage of

Immunodeficiency, microcephaly with normal intelligence, genetic damage of

Imperforate anus, genetic damage of

Imperforate oropharynx costo vetebral anomalies, genetic damage of

Impossible syndrome, genetic damage of

Inborn amino acid metabolism disorder, genetic damage of

Inborn branched chain aminoaciduria, genetic damage of

Inborn error of metabolism, genetic damage of

Inborn metabolic disorder, genetic damage of

Inborn renal aminoaciduria, genetic damage of

Inborn urea cycle disorder, genetic damage of

Incisors fused, genetic damage of

Incontinentia pigmenti, genetic damage of

Incontinentia pigmenti type 1, genetic damage of

Incontinentia pigmenti type 2, genetic damage of

Infant epilepsy with migrant focal crisis, genetic damage of

Infantile apnea, genetic damage of

Infantile axonal neuropathy, genetic damage of

Infantile dysphagia, genetic damage of

Infantile multisystem inflammatory disease, genetic damage of

Infantile myofibromatosis, genetic damage of

Infantile onset spinocerebellar ataxia, genetic damage of

Infantile recurrent chronic multifocal osteomyolitis, genetic dam-

age of
Infantile sialic acid storage disorder, genetic damage of
Infantile spasms, genetic damage of
Infantile spasms broad thumbs, genetic damage of
Infantile spinal muscular atrophy, genetic damage of
Infantile striato thalamic degeneration, genetic damage of
Inflammatory breast cancer, genetic damage of
Infundibulopelvic stenosis multicystic kidney, genetic damage of
Insensitivity to pain with anhidrosis, genetic damage of
Instability mitotic non-disjunction, genetic damage of
Insulin-resistance type B, genetic damage of
Insulinoma, genetic damage of
Intercellular cholesterol esterification disease, genetic damage of
Interferon gamma, receptor 1, deficiency, genetic damage of
Internal carotid agenesis, genetic damage of
Interstitial cystitis, genetic damage of
Interstitial pneumonia, genetic damage of
Intestinal atresia multiple, genetic damage of
Intestinal lipodystrophy, genetic damage of
Intestinal malrotation facial anomalies familial type, genetic damage of
Intestinal pseudo-obstruction chronic idiopathic, genetic damage of
Intoeing, genetic damage of
Intracranial aneurysms multiple anomaly, genetic damage of
Intracranial arterioveinous malformation, genetic damage of
Intrathoracic kidney vertebral fusion, genetic damage of
Intrauterine growth retardation mandibular malar hypoplasia, genetic damage of
Intrinsic factor, deficiency of, genetic damage of
Iophobia, genetic damage of (possible)
Iridogoniodysgenesis, dominant type, genetic damage of
Iris dysplasia hypertelorism deafness, genetic damage of

J

Jackson-Weiss syndrome, genetic damage of
Jacobs syndrome, genetic damage of
Jacobsen syndrome, genetic damage of
Jadassohn Lewandowsky syndrome, genetic damage of
Jaffer Beighton syndrome, genetic damage of
Jalili syndrome, genetic damage of
Jancar syndrome, genetic damage of
Jankovic Rivera syndrome, genetic damage of
Jansen type metaphyseal chondrodysplasia, genetic damage of
Jansky-Bielschowsky disease, genetic damage of
Japanese encephalitis, genetic damage of
Jarcho-Levin syndrome, genetic damage of
Jejunal atresia, genetic damage of
Jensen syndrome, genetic damage of
Jequier Kozlowski skeletal dysplasia, genetic damage of
Jervell Lange-Nielsen syndrome, genetic damage of
Jeune syndrome, genetic damage of
Jeune syndrome situs inversus, genetic damage of
Job syndrome, genetic damage of
Johanson Blizzard syndrome, genetic damage of
Johnson Hall Krous syndrome, genetic damage of
Johnson Munson syndrome, genetic damage of
Johnston Aarons Schelley syndrome, genetic damage of
Joint instability syndrome, genetic damage of
Joncs Hcrsh Yusk syndrome, genetic damage of
Jones syndrome, genetic damage of
Jorgenson Lenz syndrome, genetic damage of
Joseph disease, genetic damage of
Joubert syndrome, genetic damage of
Joubert syndrome bilateral chorioretinal coloboma, genetic damage of
Joubert-Boltshauser syndrome, genetic damage of

Juberg Hayward syndrome, genetic damage of
Juberg Marsidi syndrome, genetic damage of
Judge Misch Wright syndrome, genetic damage of
Jumping Frenchmen of Maine, genetic damage of
Jung Wolff Back Stahl syndrome, genetic damage of
Juvenile arthritis, genetic damage of
Juvenile cataract cerebellar atrophy myopathy mental retardation, genetic damage of
Juvenile chronic arthritis, genetic damage of
Juvenile dermatomyositis, genetic damage of
Juvenile gastrointestinal polyposis, genetic damage of
Juvenile hyaline fibromatosis, genetic damage of
Juvenile macular degeneration hypotrichosis, genetic damage of
Juvenile muscular atrophy of the distal upper limb, genetic damage of
Juvenile myoclonic epilepsy, genetic damage of
Juvenile nephronophthisis, genetic damage of
Juvenile rheumatoid arthritis, genetic damage of
Juvenile temporal arteritis, genetic damage of

K

Kabuki make up syndrome, genetic damage of
Kalam Hafeez syndrome, genetic damage of
Kaler Garrity Stern syndrome, genetic damage of
Kallikrein hypertension, genetic damage of
Kallman's syndrome, genetic damage of
Kallmann syndrome, genetic damage of
Kallmann syndrome heart disease, genetic damage of
Kallmann syndrome, type 1, X linked, genetic damage of
Kallmann syndrome, type 2, dominant, genetic damage of
Kallmann syndrome, type 3, recessive, genetic damage of
Kalyanraman syndrome, genetic damage of
Kantaputra Gorlin syndrome, genetic damage of
Kaplan Plauchu Fitch syndrome, genetic damage of
Kaplowitz Bodurtha syndrome, genetic damage of
Kaposi sarcoma, genetic damage of
Kaposiform hemangio-endothelioma, genetic damage of
Kapur Toriello syndrome, genetic damage of
Karandikar Maria Kamble syndrome, genetic damage of
Karsch Neugebauer syndrome, genetic damage of
Kartagener syndrome, genetic damage of
Kashani Strom Utley syndrome, genetic damage of
Kasznica Carlson Coppedge syndrome, genetic damage of
Katagelophobia, genetic damage of (possible)
Kathisophobia, genetic damage of (possible)
Katsantoni Papadakou Lagoyanni syndrome, genetic damage of
Katz syndrome, genetic damage of
Kaufman McKusick syndrome, genetic damage of
Kaufman oculocerebrofacial syndrome, genetic damage of
Kawasaki syndrome, genetic damage of
KBG syndrome, genetic damage of
Kearns-Sayre syndrome, genetic damage of
Keloids, genetic damage of, genetic damage of

Kennedy's disease, genetic damage of
Kennerknecht Sorgo Oberhoffer syndrome, genetic damage of
Kennerknecht Vogel syndrome, genetic damage of
Kenny Caffe syndrome, genetic damage of
Kenny syndrome, genetic damage of
Kenophobia, genetic damage of (possible)
Keratitis ichthyosis deafness syndrome, genetic damage of
Keratitis, hereditary, genetic damage of
Keratoacanthoma, genetic damage of
Keratoacanthoma familial, genetic damage of
Keratoconjunctivitis, genetic damage of
Keratoconjunctivitis sicca, genetic damage of
Keratoconus, genetic damage of
Keratoconus posticus circumscriptus, genetic damage of
Keratoderma ainhumoid and mutilans, genetic damage of
Keratoderma hypotrichosis leukonychia, genetic damage of
Keratoderma palmoplantar deafness, genetic damage of
Keratoderma palmoplantar spastic paralysis, genetic damage of
Keratoderma palmoplantaris transgrediens, genetic damage of
Keratodermia palmoplantar periorificial, genetic damage of
Keratolytic winter erythema, genetic damage of
Keratomalacia, genetic damage of
Keratosis, genetic damage of
Keratosis focal palmoplantar gingival, genetic damage of
Keratosis follicularis dwarfism cerebral atrophy, genetic damage of
Keratosis follicularis spinulosa decalvans (ichthyosis), genetic damage of
Keratosis palmoplantar-periodontopathy, genetic damage of
Keratosis palmoplantaris adenocarcinoma of the colon, genetic damage of
Keratosis palmoplantaris oesophageal colon cancer, genetic damage of
Keratosis palmoplantaris papulosa, genetic damage of

Keratosis palmoplantaris periodontopathia, genetic damage of
Keratosis palmoplantaris with corneal dystrophy, genetic damage of
Keratosis pilaris, genetic damage of
Keratosis pilaris atrophicans, genetic damage of
Keratosis, seborrheic, genetic damage of
Kernicterus, genetic damage of
Ketoaciduria mental deficiency ataxia deafness, genetic damage of
Ketotic hyperglycinemia, genetic damage of
Khalifa Graham syndrome, genetic damage of
Ki1-cell lymphoma, genetic damage of
KID syndrome, genetic damage of
Kidney and other urinary tract cancers, genetic damage of
Kienboeck disease, genetic damage of
Kikuchi disease, genetic damage of
Kimura disease, genetic damage of
Kinetophobia, genetic damage of (possible)
King-Denborough syndrome, genetic damage of
Kleeblattschaedel syndrome, genetic damage of
Klein-Waardenburg syndrome, genetic damage of
Kleine Levin syndrome, genetic damage of
Kleiner Holmes syndrome, genetic damage of
Kleptophobia, genetic damage of (possible)
Klippel Feil deformity conductive deafness absent, genetic damage of
Klippel Feil syndrome dominant type, genetic damage of
Klippel Feil syndrome recessive type, genetic damage of
Klippel Trenaunay Weber syndrome, genetic damage of
Klippel-Feil syndrome, genetic damage of
Klumpke paralysis, genetic damage of
Kluver-Bucy syndrome, genetic damage of
Kniest dysplasia, genetic damage of
Kniest like dysplasia lethal, genetic damage of

Knobloch layer syndrome, genetic damage of
Knuckle pods leuconychia sensorineural deafness, genetic damage of
Kobberling-Dunnigan syndrome, genetic damage of
Kocher-Debré-Semélaigne syndrome, genetic damage of
Kohler disease, genetic damage of
Kohlschutter Tonz syndrome, genetic damage of
Kok disease, genetic damage of
Konigsmark Knox Hussels syndrome, genetic damage of
Koone Rizzo Elias syndrome, genetic damage of
Kopophobia, genetic damage of (possible)
Korsakoff's syndrome, genetic damage of
Korula Wilson Salomon syndrome, genetic damage of
Kostmann syndrome, genetic damage of
Kosztolanyi syndrome, genetic damage of
Kotzot-Richter syndrome, genetic damage of
Koussef Nichols syndrome, genetic damage of
Kousseff syndrome, genetic damage of
Kowarski syndrome, genetic damage of
Kozlowski Brown Hardwick syndrome, genetic damage of
Kozlowski Celermajer syndrome, genetic damage of
Kozlowski Massen syndrome, genetic damage of
Kozlowski Ouvrier syndrome, genetic damage of
Kozlowski Rafinski Klicharska syndrome, genetic damage of
Kozlowski Tsuruta Taki syndrome, genetic damage of
Kozlowski Warren Fisher syndrome, genetic damage of
Kozlowski-Krajewska syndrome, genetic damage of
Krabbe leukodystrophy, genetic damage of
Krasnow Qazi syndrome, genetic damage of
Krauss Herman Holmes syndrome, genetic damage of
Krieble Bixler syndrome, genetic damage of
KTW, genetic damage of
Kufs disease, genetic damage of

Kugelberg-Welander syndrome, genetic damage of
Kumar Levick syndrome, genetic damage of
Kunze Riehm syndrome, genetic damage of
Kurczynski Casperson syndrome, genetic damage of
Kuskokwim disease, genetic damage of
Kuster Majewski Hammerstein syndrome, genetic damage of
Kuster syndrome, genetic damage of
Kuzniecky syndrome, genetic damage of
Kyasanur Forrest disease, genetic damage of
Kyphophobia, genetic damage of (possible)
Kyphosis brachyphalangy optic atrophy, genetic damage of

L

L-transposition and ccTGA, genetic damage of
Labyrinthitis syndrome, genetic damage of
Lachanophobia, genetic damage of (possible)
Lachiewicz Sibley syndrome, genetic damage of
Lacrimo-auriculo-dento-digital syndrome, genetic damage of
Lactate dehydrogenase deficiency, genetic damage of
Lactate dehydrogenase deficiency type A, genetic damage of
Lactate dehydrogenase deficiency type B, genetic damage of
Lactate dehydrogenase deficiency type C, genetic damage of
Lactic acidosis congenital infantile, genetic damage of
Lactose intolerance, genetic damage of
Ladd syndrome, genetic damage of
Ladda Zonana Ramer syndrome, genetic damage of
Lafora body disorder, genetic damage of
Lafora disease, genetic damage of
Lagophthalmia cleft lip palate, genetic damage of
Lambdoid synostosis familial, genetic damage of
Lambert syndrome, genetic damage of
Lambert-Eaton Myasthenic Syndrome (Lambert-Eaton paraneo-
plastic cerebellar degeneration), genetic damage of
Lambert-Eaton syndrome, genetic damage of
Lamellar ichtyosis, genetic damage of
Lamellar recessive ichthyosis, genetic damage of
Landau-Kleffner syndrome, genetic damage of
Landing disease, genetic damage of
Landouzy-Dejerine muscular dystrophy, genetic damage of
Landy Donnai syndrome, genetic damage of
Langdon Down, genetic damage of
Langer Nishino Yamaguchi syndrome, genetic damage of
Langer-Giedion syndrome, genetic damage of
Langerhans cell granulomatosis, genetic damage of
Langerhans cell histiocytosis, genetic damage of

Laparoschisis, genetic damage of
Laplane Fontaine Lagardere syndrome, genetic damage of
Large B cell diffuse lymphoma, genetic damage of
Laron syndrome, genetic damage of
Laron-type dwarfism, genetic damage of
Larsen like osseous dysplasia dwarfism, genetic damage of
Larsen like syndrome lethal type, genetic damage of
Larsen syndrome, genetic damage of
Larsen syndrome craniosynostosis, genetic damage of
Larsen syndrome, dominant type, genetic damage of
Larsen syndrome, recessive type, genetic damage of
Laryngeal abductor paralysis, genetic damage of
Laryngeal abductor paralysis mental retardation, genetic damage of
Laryngeal carcinoma, genetic damage of
Laryngeal cleft, genetic damage of
Laryngeal neoplasm, genetic damage of
Laryngeal papillomatosis, genetic damage of (possible)
Laryngeal web congenital heart disease short stature, genetic damage of
Laryngocele, genetic damage of
Laryngomalacia, genetic damage of
Laryngomalacia dominant congenital, genetic damage of
Laryngotracheoesophageal cleft pulmonary hypoplasia, genetic damage of
Larynx atresia, genetic damage of
Lassueur-Graham-Little syndrome, genetic damage of
Late onset dominant cone dystrophy, genetic damage of
Lateral body wall defect, genetic damage of
Laterality defects dominant, genetic damage of
Lattice corneal dystrophy type 2, genetic damage of
Launois-Bensaude adenolipomatosis, genetic damage of
Laurence Prosser Rocker syndrome, genetic damage of
Laurence-Moon-Bardet-Biedl syndrome, genetic damage of

Laurin Sandrow syndrome, genetic damage of
Laxova Brown Hogan syndrome, genetic damage of
LBWC—amniotic bands, genetic damage of
LBWD syndrome, genetic damage of
LCHAD deficiency, genetic damage of
LCHAD long-chain 3-hydroxyacyl-CoA dehydrogenase deficiency
(G1528C mutation), genetic damage of
Leao Ribeiro Da Silva syndrome, genetic damage of
Learman syndrome, genetic damage of
Leber congenital amaurosis, genetic damage of
Leber hereditary optic neuropathy, genetic damage of
Leber miliary aneurysm, genetic damage of
Leber optic atrophy, genetic damage of
Lecithin cholesterol acyltransferase deficiency, genetic damage of
Ledderhose disease, genetic damage of
Lee Root Fenske syndrome, genetic damage of
Left ventricle-aorta tunnel, genetic damage of
Leg absence deformity cataract, genetic damage of
Legg-Calvé-Perthes syndrome, genetic damage of
Lehman syndrome, genetic damage of
Leichtman Wood Rohn syndrome, genetic damage of
Leifer Lai Buyse syndrome, genetic damage of
Leigh disease, genetic damage of
Leiner disease (complement component 5 deficiency), genetic
damage of
Leiomyoma, genetic damage of
Leiomyomatosis familial, genetic damage of
Leiomyomatosis of oesophagus cataract hematuria, genetic damage
of
Leiomyosarcoma, genetic damage of
Leipala Kaitila syndrome, genetic damage of
Leisti Hollister Rimoin syndrome, genetic damage of
Lemierre's syndrome, genetic damage of

Lennox-Gastaut syndrome, genetic damage of
Lentiginosis in context of NF, genetic damage of
Lenz Majewski hyperostotic dwarfism, genetic damage of
Lenz microphthalmia syndrome, genetic damage of
Leprechaunism, genetic damage of
Leprophobia, genetic damage of (possible)
Leptomeningeal capillary—venous angiomatosis, genetic damage of
Leri pleonosteosis, genetic damage of
Leri-Weil syndrome, genetic damage of
Leshima Koeda Inagaki syndrome, genetic damage of
Lethal chondrodysplasia Moerman type, genetic damage of
Lethal chondrodysplasia Seller type, genetic damage of
Lethal congenital contracture syndrome, genetic damage of
Letterer-Siwe disease, genetic damage of
Leucinosis, genetic damage of
Leukemia, genetic damage of
Leukemia subleukemic, genetic damage of
Leukemia, B-Cell, chronic, genetic damage of
Leukemia, Myeloid, genetic damage of
Leukemia, T-Cell, chronic, genetic damage of
Leukocyte adhesion deficiency syndrome, genetic damage of
Leukocyte adhesion deficiency type 2, genetic damage of
Leukocytoclastic angiitis, genetic damage of
Leukodystrophy, genetic damage of
Leukodystrophy reunion type, genetic damage of
Leukodystrophy, globoid cell, genetic damage of
Leukodystrophy, metachromatic, genetic damage of
Leukoencephalopathy palmoplantar keratoderma, genetic damage of
Leukomalacia, genetic damage of
Leukomelanoderma mental redardation hypotrichosis, genetic damage of

Leukophobia, genetic damage of (possible)
Leukoplakia, genetic damage of
Levator syndrome, genetic damage of
Levic Stefanovic Nikolic syndrome, genetic damage of
Levine Crichley syndrome, genetic damage of
Levy Hollister syndrome, genetic damage of
Lewandowski Kikolich syndrome, genetic damage of
Lewis Pashayan syndrome, genetic damage of
Lewy body dementia, genetic damage of
Lewy body disease, genetic damage of
Leydig cells hypoplasia, genetic damage of
LGCR, genetic damage of
LGS, genetic damage of
Lhermitte-Duclos disease, genetic damage of
Li-Fraumeni syndrome, genetic damage of
Lichen myxedematosus, genetic damage of
Lichen planus, genetic damage of
Lichen planus follicularis, genetic damage of
Lichen sclerosis et atrophicus, genetic damage of
Lichstenstein syndrome, genetic damage of
Lida Kannari syndrome, genetic damage of
Liddle syndrome, genetic damage of
Light chain disease, genetic damage of
Ligyrophobia, genetic damage of (possible)
Limb deficiencies distal micrognathia, genetic damage of
Limb dystonia, genetic damage of
Limb reduction defect, genetic damage of
Limb scalp and skull defects, genetic damage of
Limb transversal defect cardiac anomaly, genetic damage of
Limb-body wall complex, genetic damage of
Limb-girdle muscular dystrophy, genetic damage of
Limnophobia, genetic damage of (possible)
Lindsay Burn syndrome, genetic damage of

Lindstrom syndrome, genetic damage of
Linear hamartoma syndrome, genetic damage of
Linear nevus syndrome, genetic damage of
Linonophobia, genetic damage of (possible)
Lip lit syndrome, genetic damage of
Lipid storage myopathy, genetic damage of
Lipidosis with triglycerid storage disease, genetic damage of
Lipoamide dehydrogenase deficiency, genetic damage of
Lipoatrophic diabetes, genetic damage of
Lipodystrophy, genetic damage of
Lipodystrophy Rieger anomaly diabetes, genetic damage of
Lipogranulomatosis, genetic damage of
Lipoid congenital adrenal hyperplasia, genetic damage of
Lipoid proteinosis of Urbach and Wiethe, genetic damage of
Lipomatosis central non-encapsulated, genetic damage of
Lipomatosis familial benign cervical, genetic damage of
Lipomatosis of pancreas, congenital, genetic damage of
Lipomucopolysaccharidosis, genetic damage of
Lipoprotein disorder, genetic damage of
Liposarcoma, genetic damage of
Lisker Garcia Ramos syndrome, genetic damage of
Lison Kornbrut Feinstein syndrome, genetic damage of
Lissencephaly, genetic damage of
Lissencephaly immunodeficiency, genetic damage of
Lissencephaly syndrome type 1, genetic damage of
Lissencephaly syndrome type 2, genetic damage of
Lissencephaly, isolated, genetic damage of
Liticaphobia, genetic damage of (possible)
Liver cirrhosis, genetic damage of
Liver neoplasms, genetic damage of
Lobar atrophy of brain, genetic damage of
Lobstein disease, genetic damage of
Localized epiphyseal dysplasia, genetic damage of

Lockwood Feingold syndrome, genetic damage of
Loffredo Cennamo Cecio syndrome, genetic damage of
Logic syndrome, genetic damage of
Loiasis, genetic damage of
Loin pain hematuria syndrome, genetic damage of
Long QT Syndrome, genetic damage of
Long QT syndrome type 1, genetic damage of
Long QT syndrome type 2, genetic damage of
Long QT syndrome type 3, genetic damage of
Loose anagen hair syndrome, genetic damage of
Loose anagene syndrome, genetic damage of
Lopes Gorlin syndrome, genetic damage of
Lopes Marques de Faria syndrome, genetic damage of
Lopez Hernandez syndrome, genetic damage of
Lou Gehrig's disease, genetic damage of
Louis Bar syndrome, genetic damage of
Low birth weight dwarfism dysgammaglobulinemia, genetic damage of
Lowe Kohn Cohen syndrome, genetic damage of
Lowe oculocerebrorenal syndrome, genetic damage of
Lowe syndrome, genetic damage of
Lower limb anomaly ureteral obstruction, genetic damage of
Lower limb deficiency hypospadias, genetic damage of
Lower limb partial duplication renal agenesis, genetic damage of
Lower mesodermal defects, genetic damage of
Lowry Maclean syndrome, genetic damage of
Lowry syndrome, genetic damage of
Lowry Wood syndrome, genetic damage of
Lowry Yong syndrome, genetic damage of
LSA, genetic damage of
Lubani Al Saleh Teebi syndrome, genetic damage of
Lubinsky syndrome, genetic damage of
Lucey Driscoll syndrome, genetic damage of

Lucky Gelehrter syndrome, genetic damage of
Luiphobia, genetic damage of (possible)
Lujan-Fryns syndrome, genetic damage of
Lumbar malsegmentation short stature, genetic damage of
Lundberg syndrome, genetic damage of
Lung agenesis heart defect thumb anomalies, genetic damage of
Lung cancer, genetic damage of
Lung herniation congenital defect of sternem, genetic damage of
Lung neoplasm, genetic damage of
Lupus, genetic damage of
Lupus anticoagulant, familial, genetic damage of
Lurie Kletsky syndrome, genetic damage of
Luteinizing hormone releasing hormone, deficiency of with ataxia, genetic damage of
Lutz Richner Landolt syndrome, genetic damage of
Lutz-Lewandowsky epidermodysplasia verruciformis, genetic damage of
Lyell's syndrome, genetic damage of
Lygophobia, genetic damage of (possible)
Lymph node neoplasm, genetic damage of
Lymphadenopathy, angioimmunoblastic with dysproteinemia, genetic damage of
Lymphangiectasies lymphoedema type Hennekam type, genetic damage of
Lymphangiectasis, genetic damage of
Lymphangioleiomyomatosis, genetic damage of
Lymphangiomatosis, pulmonary, genetic damage of
Lymphangiomyomatosis, genetic damage of
Lymphatic filariasis, genetic damage of
Lymphatic neoplasm, genetic damage of
Lymphedema, genetic damage of
Lymphedema distichiasis, genetic damage of
Lymphedema hereditary type 1, genetic damage of

Lymphedema hereditary type 2, genetic damage of
Lymphedema ptosis, genetic damage of
Lymphedema, congenital, genetic damage of
Lymphoblastic lymphoma, genetic damage of
Lymphocytes; reduced or absent, genetic damage of
Lymphocytic colitis, genetic damage of
Lymphocytic infiltrate of Jessner, genetic damage of
Lymphocytic vasculitis, genetic damage of
Lymphoid hamartoma, genetic damage of
Lymphoma, genetic damage of
Lymphoma, gastric non Hodgkins type, genetic damage of
Lymphoma, large-cell, genetic damage of
Lymphoma, large-cell, immunoblastic, genetic damage of
Lymphoma, small cleaved-cell, diffuse, genetic damage of
Lymphoma, small cleaved-cell, follicular, genetic damage of
Lymphomatoid granulomatosis, genetic damage of
Lymphomatoid papulosis (LyP), genetic damage of
Lymphomatous thyroiditis, genetic damage of
Lymphosarcoma, genetic damage of
Lynch Lee Murday syndrome, genetic damage of
Lynch syndrome, genetic damage of
Lynch-Bushby syndrome, genetic damage of
Lyngstadaas syndrome, genetic damage of
LyP (lymphomatoid papulosis), genetic damage of
Lysine alpha-ketoglutarate reductase deficiency, genetic damage of
Lysinuric protein intolerance, genetic damage of
Lysosomal alpha-D-mannosidase deficiency, genetic damage of
Lysosomal beta-mannosidase deficiency, genetic damage of
Lysosomal disorders, genetic damage of
Lysosomal glycogen storage disease with normal acid maltase activity, genetic damage of

M

Maaswinkel Mooij Stokvis Brantsma syndrome, genetic damage of

Mac Dermot Patton Williams syndrome, genetic damage of

Mac Dermot Winter syndrome, genetic damage of

Machado-Joseph disease (MJD), genetic damage of

Macias Flores Garcia Cruz Rivera syndrome, genetic damage of

Mackay Shek Carr syndrome, genetic damage of

Macleod Fraser syndrome, genetic damage of

Macrocephaly cutis marmorata telangiectatica, genetic damage of

Macrocephaly dominant type, genetic damage of

Macrocephaly mental retardation facial dysmorphism, genetic damage of

Macrocephaly mesodermal hamartoma spectrum, genetic damage of

Macrocephaly mesomelic arms talipes, genetic damage of

Macrocephaly pigmentation large hands feet, genetic damage of

Macrocephaly short stature paraplegia, genetic damage of

Macrodactyly of the foot, genetic damage of

Macroepiphyseal dysplasia Mcalister Coe type, genetic damage of

Macroglobulinemia, genetic damage of

Macroglossia dominant, genetic damage of

Macroglossia exomphalos gigantism, genetic damage of

Macrogyria pseudobulbar palsy, genetic damage of

Macrophagic myofasciitis, genetic damage of

Macrosomia developmental delay dysmorphism, genetic damage of

Macrosomia microphthalmia cleft palate, genetic damage of

Macrothrombocytopenia progressive deafness, genetic damage of

Macrothrombocytopenia with leukocyte inclusions, genetic damage of

Macular corneal dystrophy, genetic damage of

Macular degeneration, genetic damage of

Macular degeneration juvenile, genetic damage of

Macular degeneration, age-related, genetic damage of
Macular degeneration, polymorphic, genetic damage of
Macular dystrophy, vitelliform, genetic damage of
Macules hereditary congenital hypopigmented and hyperpigmented, genetic damage of
Madelung's disease, genetic damage of
Madokoro Ohdo Sonoda syndrome, genetic damage of
Maffucci syndrome, genetic damage of
Mageirocophobia, genetic damage of (possible)
Maghazaji syndrome, genetic damage of
Magnesium defect in renal tubular transport of, genetic damage of
Magnesium wasting renal, genetic damage of
Mal de Debarquement, genetic damage of
Malakoplakia, genetic damage of
Male pseudohermaphroditism due to 17-beta-hydroxysteroid dehydrogenase deficiency, genetic damage of
Male pseudohermaphroditism due to 5-alpha-reductase 2 deficiency, genetic damage of
Male pseudohermaphroditism due to androgen insensitivity, genetic damage of
Male pseudohermaphroditism due to defective LH molecule, genetic damage of
Male Turner syndrome, genetic damage of
Malformations in neuronal migration, genetic damage of
Malignant astrocytoma, genetic damage of
Malignant fibrous histiocytoma, genetic damage of
Malignant germ cell tumor, genetic damage of
Malignant hyperthermia, genetic damage of
Malignant hyperthermia arthrogryposis torticollis, genetic damage of
Malignant hyperthermia susceptibility, genetic damage of
Malignant hyperthermia susceptibility type 1, genetic damage of
Malignant hyperthermia susceptibility type 2, genetic damage of

Malignant hyperthermia susceptibility type 3, genetic damage of
Malignant hyperthermia susceptibility type 4, genetic damage of
Malignant hyperthermia susceptibility type 5, genetic damage of
Malignant hyperthermia susceptibility type 6, genetic damage of
Malignant mesenchymal tumor, genetic damage of
Malignant mixed Mullerian tumor, genetic damage of
Malignant paroxysmal ventricular tachycardia, genetic damage of
Mallory-Weiss syndrome, genetic damage of
Malonic aciduria, genetic damage of
Malonyl-CoA decarboxylase deficiency, genetic damage of
Malouf syndrome, genetic damage of
Mandibuloacral dysplasia, genetic damage of
Mandibulofacial dysostosis deafness postaxial polydactly, genetic damage of
Manic depression, bipolar, genetic damage of
Manic-depressive psychosis, genetic types, genetic damage of
Mannosidosis, genetic damage of
Manouvrier syndrome, genetic damage of
Mantle cell lymphoma, genetic damage of
Maple syrup urine disease, genetic damage of
Marashi Gorlin syndrome, genetic damage of
Marchiafava Bignami disease, genetic damage of
Marchiafava-Micheli disease, genetic damage of
Marcus Gunn phenomenon, genetic damage of
Marden Walker like syndrome, genetic damage of
Marden-Walker syndrome, genetic damage of
Marek disease, genetic damage of
Marfan syndrome, genetic damage of
Marfan Syndrome type I, genetic damage of
Marfan Syndrome type II, genetic damage of
Marfan Syndrome type III, genetic damage of
Marfan Syndrome type IV, genetic damage of
Marfan Syndrome type V, genetic damage of

Marfan-Like syndrome, genetic damage of
Marfan-like syndrome, Boileau type, genetic damage of
Marfanoid build spondylolisthesis constricted pelvis, genetic damage of
Marfanoid craniosynostosis syndrome, genetic damage of
Marfanoid hypermobility, genetic damage of
Marfanoid mental retardation syndrome autosomal, genetic damage of
Marginal glioneuronal heterotopia, genetic damage of
Marie type ataxia, genetic damage of
Marie Unna congenital hypotrichosis, genetic damage of
Marineaco-Sjogren syndrome, genetic damage of
Marinesco Sjogren like syndrome, genetic damage of
Marion Mayers syndrome, genetic damage of
Markel Vikkula Mulliken syndrome, genetic damage of
Marles Greenberg Persaud syndrome, genetic damage of
Maroteaux Cohen Solal Bonaventure syndrome, genetic damage of
Maroteaux Fonfria syndrome, genetic damage of
Maroteaux Le Merrer Bensahel syndrome, genetic damage of
Maroteaux Stanescu Cousin syndrome, genetic damage of
Maroteaux Verloes Stanescu syndrome, genetic damage of
Maroteaux-Lamy syndrome, genetic damage of
Marphanoid syndrome type De Silva, genetic damage of
Marsden Nyhan Sakati syndrome, genetic damage of
Marshall syndrome, genetic damage of
Marshall-Smith syndrome, genetic damage of
Martin Bell syndrome, genetic damage of
Martinez Monasterio Pinheiro syndrome, genetic damage of
Martsolf Reed Hunter syndrome, genetic damage of
Martsolf syndrome, genetic damage of
MASA syndrome, genetic damage of
Massa Casaer Ceulemans syndrome, genetic damage of
Mast cell disease, genetic damage of

Mastigophobia, genetic damage of (possible)
Mastocytosis, genetic damage of
Mastocytosis, short stature, hearing loss, genetic damage of
Mastroiacovo De Rosa Satta syndrome, genetic damage of
Mastroiacovo Gambi Segni syndrome, genetic damage of
MAT deficiency, genetic damage of
Maternal hyperphenylalaninemia, genetic damage of
Maternally inherited diabetes and deafness, genetic damage of
Mathieu De Broca Bony syndrome, genetic damage of
Matsoukas Liarikos Giannika syndrome, genetic damage of
Matthew-Wood syndrome, genetic damage of
Maturity onset diabetes of the young, genetic damage of
Maumenee syndrome, genetic damage of
Maxillary double lip, genetic damage of
Maxillofacial dysostosis, genetic damage of
Maxillonasal dysplasia, Binder type, genetic damage of
May-Hegglin anomaly, genetic damage of
Mayer Rokitanski Kuster syndrome, genetic damage of
McAlister Crane syndrome, genetic damage of
McArdle disease, genetic damage of
McCallum Macadam Johnston syndrome, genetic damage of
McCune-Albright syndrome, genetic damage of
McDonough syndrome, genetic damage of
McDowall syndrome, genetic damage of
McGillivray syndrome, genetic damage of
McKusick Kaufman syndrome, genetic damage of
McKusick type metaphyseal chondrodysplasia, genetic damage of
McLain Debakian syndrome, genetic damage of
McPherson Clemens syndrome, genetic damage of
McPherson Robertson Cammarano syndrome, genetic damage of
Meacham Winn Culler syndrome, genetic damage of
Meadows syndrome, genetic damage of
Meckel like syndrome, genetic damage of

Meckel syndrome, genetic damage of
Medeira Dennis Donnai syndrome, genetic damage of
Median cleft lip corpus callosum lipoma skin polyps, genetic damage of
Median nodule of the upper lip, genetic damage of
Mediastinal endodermal sinus tumors, genetic damage of
Medium-chain Acyl-CoA dehydrogenase deficiency, genetic damage of
Medrano Roldan syndrome, genetic damage of
Medullary cystic disease, genetic damage of
Medullary thyroid carcinoma, genetic damage of
Medulloblastoma, genetic damage of
Megacystis microcolon intestinal hypoperistalsis syndrome, genetic damage of
Megaduodenum and/or megacystis, genetic damage of
Megaepiphyseal dwarfism, genetic damage of
Megalencephalic leukodystrophy, genetic damage of
Megalencephaly-cystic leukodystrophy, genetic damage of
Megaloblastic anemia, genetic damage of
Megalocornea mental retardation syndrome, genetic damage of
Megalocytic interstitial nephritis, genetic damage of
Mehes syndrome, genetic damage of
Mehta Lewis Patton syndrome, genetic damage of
Meier Blumberg Imahorn syndrome, genetic damage of
Meier Gorlin syndrome, genetic damage of
Meier Rotschild syndrome, genetic damage of
Meige syndrome, genetic damage of
Meigel disease, genetic damage of
Meinecke Pepper syndrome, genetic damage of
Meinecke syndrome, genetic damage of
Melanoma type 1, genetic damage of
Melanoma type 2, genetic damage of
Melanoma, familial, genetic damage of

Melanoma, malignant, genetic damage of
Melanosis neurocutaneous, genetic damage of
MELAS, genetic damage of
Meleda disease, genetic damage of
Melhem Fahl syndrome, genetic damage of
Melkersson-Rosenthal syndrome, genetic damage of
Melnick-Needles osteodysplasty, genetic damage of
Melnick-Needles syndrome, genetic damage of
Melophobia, genetic damage of (possible)
Mendelian susceptibility to atypical mycobacteria, genetic damage of
Menetrier's disease, genetic damage of
Mengel Konigsmark syndrome, genetic damage of
Meniere's disease, genetic damage of
Meningeal angiomatosis cleft hypoplastic left heart, genetic damage of
Meningioma, genetic damage of
Meningioma 1, genetic damage of
Meningocele, genetic damage of
Meningoencephalocele, genetic damage of
Meningoencephalocele-arthrogryposis-hypoplastic thumb, genetic damage of
Meningomyelocele, genetic damage of
Menkes kinky hair syndrome, genetic damage of
Menophobia, genetic damage of (possible)
Mental deficiency-epilepsy-endocrine disorders, genetic damage of
Mental mixed retardation deafnes clubbed digits, genetic damage of
Mental retardatio-polydactyly-uncombable hair, genetic damage of
Mental retardation, genetic damage of
Mental retardation anophthalmia craniosynostosis, genetic damage of
Mental retardation arachnodactyly hypotonia telangiectasia,

genetic damage of

Mental retardation athetosis microphthalmia, genetic damage of

Mental retardation blepharophimosis obesity web neck, genetic damage of

Mental retardation Buenos Aires type, genetic damage of

Mental retardation cataracts calcified pinnae myopathy, genetic damage of

Mental retardation coloboma slimness, genetic damage of

Mental retardation contractural arachnodactyly, genetic damage of

Mental retardation dysmorphism hypogonadism diabetes, genetic damage of

Mental retardation epilepsy, genetic damage of

Mental retardation epilepsy bulbous nose, genetic damage of

Mental retardation gynecomastia obesity X linked, genetic damage of

Mental retardation hip luxation G6PD variant, genetic damage of

Mental retardation hypocupremia hypobetalipoproteinemia, genetic damage of

Mental retardation hypotonia skin hyperpigmentation, genetic damage of

Mental retardation macrocephaly coarse facies hypotonia, genetic damage of

Mental retardation microcephaly phalangeal facial, genetic damage of

Mental retardation microcephaly unusual facies, genetic damage of

Mental retardation Mietens Weber type, genetic damage of

Mental retardation multiple nevi, genetic damage of

Mental retardation myopathy short stature endocrine defect, genetic damage of

Mental retardation nasal hypoplasia obesity genital hypoplasia, genetic damage of

Mental retardation nasal papillomata, genetic damage of

Mental retardation osteosclerosis, genetic damage of

Mental retardation progressive spasticity, genetic damage of

Mental retardation psychosis macroorchidism, genetic damage of

Mental retardation short broad thumbs, genetic damage of

Mental retardation short stature absent phalanges, genetic damage of

Mental retardation short stature Bombay phenotype, genetic damage of

Mental retardation short stature cleft palate unusual facies, genetic damage of

Mental retardation short stature deafness genital, genetic damage of

Mental retardation short stature hand contractures genital anomalies, genetic damage of

Mental retardation short stature heart and skeletal anomalies, genetic damage of

Mental retardation short stature hypertelorism, genetic damage of

Mental retardation short stature microcephaly eye, genetic damage of

Mental retardation short stature ocular and articular anomalies, genetic damage of

Mental retardation short stature scoliosis, genetic damage of

Mental retardation short stature unusual facies, genetic damage of

Mental retardation short stature wedge shaped epiphyses, genetic damage of

Mental retardation skeletal dysplasia abducens palsy, genetic damage of

Mental retardation Smith Fineman Myers type, genetic damage of

Mental retardation spasticity ectrodactyly, genetic damage of

Mental retardation unusual facies, genetic damage of

Mental retardation unusual facies Ampola type, genetic damage of

Mental retardation unusual facies Davis Lafer type, genetic damage of

Mental retardation unusual facies talipes hand anomalies, genetic

damage of

Mental retardation Wolff type, genetic damage of

Mental retardation X linked Atkin type, genetic damage of

Mental retardation X linked borderline Maoa metabolism anomaly, genetic damage of

Mental retardation X linked Brunner type, genetic damage of

Mental retardation X linked dysmorphism, genetic damage of

Mental retardation X linked dystonia dysarthria, genetic damage of

Mental retardation X linked severe Gustavson type, genetic damage of

Mental retardation X linked short stature obesity, genetic damage of

Mental retardation X linked Tranebjaerg type seizures psoriasis, genetic damage of

Mental retardation, X linked, Juberg-Marsidi type, genetic damage of

Mental retardation, X linked, Marfanoid habitus, genetic damage of

Mental retardation, X linked, nonspecific, genetic damage of

Mental retardation-unusual facies-intrauterine growth, genetic damage of

Meretoja syndrome, genetic damage of

Merkle tumors, genetic damage of

Merlob Grunebaum Reisner syndrome, genetic damage of

Merlob syndrome, genetic damage of

Mesangial sclerosis, diffuse, genetic damage of

Mesenteric panniculitis, genetic damage of

Mesodermal defects lower type, genetic damage of

Mesomelia, genetic damage of

Mesomelia radial hypoplasia bifid thumb unusual facies, genetic damage of

Mesomelia synostoses, genetic damage of

Mesomelic dwarfism cleft palate camptodactyly, genetic damage of

Mesomelic dwarfism Langer type, genetic damage of
Mesomelic dwarfism Nievergelt type, genetic damage of
Mesomelic dwarfism Reinhardt Pfeiffer type, genetic damage of
Mesomelic dysplasia skin dimples, genetic damage of
Mesomelic dysplasia Thai type, genetic damage of
Mesomelic syndrome Pfeiffer type, genetic damage of
Metabolic disorder, genetic damage of
Metacarpals 4 and 5 fusion, genetic damage of
Metachondromatosis, genetic damage of
Metachromatic leukodystrophy, genetic damage of
Metageria, genetic damage of
Metaphyseal anadysplasia, genetic damage of
Metaphyseal chondrodysplasia Schmid type, genetic damage of
Metaphyseal chondrodysplasia Spahr type, genetic damage of
Metaphyseal chondrodysplasia, McKusick type, genetic damage of
Metaphyseal chondrodysplasia, others, genetic damage of
Metaphyseal dysostosis mental retardation conductive deafness, genetic damage of
Metaphyseal dysplasia maxillary hypoplasia brachydactyly, genetic damage of
Metaphyseal dysplasia Pyle type, genetic damage of
Metastatic insulinoma, genetic damage of
Metatarsus adductus, genetic damage of
Metathesiophobia, genetic damage of (possible)
Metatrophic dysplasia, genetic damage of
Metatropic dwarfism, genetic damage of
Metatropic dysplasia 1, genetic damage of
Methylcobalamin deficiency cbl G type, genetic damage of
Methylcobalamin deficiency, cbl E complementation type, genetic damage of
Methylenetetrahydrofolate reductase deficiency, genetic damage of
Methylmalonic acidemia, genetic damage of
Methylmalonic acidemia with homocystinuria, genetic damage of

Methylmalonic aciduria microcephaly cataract, genetic damage of
Methylmalonicacidemia with homocystinuria, cbl D, genetic damage of
Methylmalonicaciduria with homocystinuria, cbl F, genetic damage of
Methylmalonicaciduria, vitamin B12 unresponsive, mut-0, genetic damage of
Methylmalonyl-Coenzyme A mutase deficiency, genetic damage of
Mevalonate kinase deficiency, genetic damage of
Mevalonicaciduria, genetic damage of
Michelin tire baby syndrome, genetic damage of
Michels Caskey syndrome, genetic damage of
Michels syndrome, genetic damage of
Mickleson syndrome, genetic damage of
Micrencephaly corpus callosum agenesis, genetic damage of
Micrencephaly olivopontocerebellar hypoplasia, genetic damage of
Micro syndrome, genetic damage of
Microbrachycephaly ptosis cleft lip, genetic damage of
Microcephalic osteodysplastic primordial dwarfism, genetic damage of
Microcephalic primordial dwarfism, genetic damage of
Microcephalic primordial dwarfism Toriello type, genetic damage of
Microcephaly, genetic damage of
Microcephaly albinism digital anomalies syndrome, genetic damage of
Microcephaly autosomal dominant, genetic damage of
Microcephaly brachydactyly kyphoscoliosis, genetic damage of
Microcephaly brain defect spasticity hypernatremia, genetic damage of
Microcephaly cardiac defect lung malsegmentation, genetic damage of
Microcephaly cardiomyopathy, genetic damage of

Microcephaly cervical spine fusion anomalies, genetic damage of
Microcephaly chorioretinopathy recessive form, genetic damage of
Microcephaly cleft palate autosomal dominant, genetic damage of
Microcephaly deafness syndrome, genetic damage of
Microcephaly developmental delay pancytopenia, genetic damage of
Microcephaly facial clefting preaxial polydactyly, genetic damage of
Microcephaly glomerulonephritis Marfanoid habitus, genetic damage of
Microcephaly hiatus hernia nephrotic syndrome, genetic damage of
Microcephaly hypergonadotropic hypogonadism short stature, genetic damage of
Microcephaly immunodeficiency and chromosomal instabilty, genetic damage of
Microcephaly immunodeficiency lymphoreticuloma, genetic damage of
Microcephaly intracranial calcification, genetic damage of
Microcephaly lymphoedema chorioretinal dysplasia, genetic damage of
Microcephaly lymphoedema syndrome, genetic damage of
Microcephaly mental retardation retinopathy, genetic damage of
Microcephaly mental retardation spasticity epilepsy, genetic damage of
Microcephaly mesobrachyphalangy tracheoesophageal fistula syndrome, genetic damage of
Microcephaly microcornea syndrome Seemanova type, genetic damage of
Microcephaly micropenis convulsions, genetic damage of
Microcephaly microphthalmos blindness, genetic damage of
Microcephaly nonsyndromal, genetic damage of
Microcephaly pontocerebellar hypoplasia dyskinesia, genetic dam-

age of
Microcephaly seizures mental retardation heart disorders, genetic damage of
Microcephaly sparse hair mental retardation seizures, genetic damage of
Microcephaly syndactyly brachymesophalangy, genetic damage of
Microcephaly with chorioretinopathy, genetic damage of
Microcephaly with chorioretinopathy, autosomal dominant form, genetic damage of
Microcephaly with normal intelligence, immunodeficiency, genetic damage of
Microcephaly, primary autosomal recessive, genetic damage of
Microcoria, congenital, genetic damage of
Microcornea corectopia macular hypoplasia, genetic damage of
Microcornea glaucoma absent frontal sinuses, genetic damage of
Microdeletion 22 q11, genetic damage of
Microdontia hypodontia short stature, genetic damage of
Microencephaly, genetic damage of
Microgastria limb reduction defect, genetic damage of
Microgastria short stature diabetes, genetic damage of
Micromelic dwarfism Fryns type, genetic damage of
Micromelic dysplasia dislocation of radius, genetic damage of
Microphobia, genetic damage of (possible)
Microphtalmos bilateral colobomatous orbital cyst, genetic damage of
Microphthalmia, genetic damage of
Microphthalmia camptodactyly mental retardation, genetic damage of
Microphthalmia cataract, genetic damage of
Microphthalmia diaphragmatic hernia Fallot, genetic damage of
Microphthalmia mental deficiency, genetic damage of
Microphthalmia microtia fetal akinesia, genetic damage of
Microphthalmia, Lentz type, genetic damage of

Microphthalmos, microcornea, and sclerocornea, genetic damage of

Microscopic polyangiitis, genetic damage of

Microsomia hemifacial radial defects, genetic damage of

Microspherophakia metaphyseal dysplasia, genetic damage of

Microtia meatal atresia conductive deafness, genetic damage of

Microvillus inclusion disease, genetic damage of

Miculicz syndrome, genetic damage of

Midas syndrome, genetic damage of

Midline cleft of lower lip, genetic damage of

Midline defects autosomal type, genetic damage of

Midline defects recessive type, genetic damage of

Midline developmental field defects, genetic damage of

Midline field defects, genetic damage of

Midline lethal granuloma, genetic damage of

Mietens syndrome, genetic damage of

Mievis Verellen Dumoulin syndrome, genetic damage of

Mikati Najjar Sahli syndrome, genetic damage of

Mikulicz syndrome, genetic damage of

Miller Fisher syndrome, genetic damage of

Miller syndrome, genetic damage of

Miller-Dieker syndrome, genetic damage of

Milner Khallouf Gibson syndrome, genetic damage of

MILS syndrome, genetic damage of

Minkowski-Chauffard disease, genetic damage of

Miosis, congenital, genetic damage of

Mirror hands feet nasal defects, genetic damage of

Mirror polydactyly segmentation and limbs defects, genetic damage of

Misophobia, genetic damage of (possible)

Mitochondrial acetoacetyl-CoA thiolase deficiency, genetic damage of

Mitochondrial cytopathy (generic term), genetic damage of

Mitochondrial diseases of nuclear origin, genetic damage of
Mitochondrial diseases, clinically undefinite, genetic damage of
Mitochondrial encephalomyopathy aminoacidopathy, genetic damage of
Mitochondrial genetic disorders, genetic damage of
Mitochondrial myopathy lactic acidosis, genetic damage of
Mitochondrial myopathy-encephalopathy-lactic acidosis, genetic damage of
Mitochondrial PEPCK deficiency, genetic damage of
Mitochondrial trifunctional protein deficiency, genetic damage of
Mitral atresia, genetic damage of
Mitral regurgitation deafness skeletal anomalies, genetic damage of
Mitral valve prolapse, genetic damage of
Mitral valve prolapse, familial, autosomal dominant, genetic damage of
Mitral valve prolapse, familial, X linked, genetic damage of
Miura syndrome, genetic damage of
Mixed connective tissue disease, genetic damage of
Mixed Mullerian tumor, genetic damage of
Mixed sclerosing bone dystrophy, genetic damage of
MLS syndrome, genetic damage of
MMEP syndrome, genetic damage of
MMT syndrome, genetic damage of
MN1, genetic damage of
MNGIE syndrome, genetic damage of
MODY syndrome, genetic damage of
Moebius axonal neuropathy hypogonadism, genetic damage of
Moebius syndrome, genetic damage of
Moerman Vandenberghe Fryns syndrome, genetic damage of
Moeschler Clarren syndrome, genetic damage of
Mohr syndrome, genetic damage of
Molarization of anterior teeth deafness, genetic damage of
Mollica Pavone Antener syndrome, genetic damage of

Moloney syndrome, genetic damage of
Molybdenum cofactor deficiency, genetic damage of
Momo syndrome, genetic damage of
Mondini dysplasia, genetic damage of
Mondor's disease, genetic damage of
Monilethrix, genetic damage of
Monoamine oxidase A deficiency, genetic damage of
Monoclonal gammopathy of undetermined significance, genetic damage of
Monodactyly tetramelic, genetic damage of
Mononen Karnes Senac syndrome, genetic damage of
Mononeuritis multiplex, genetic damage of
Monosomy 10p, genetic damage of
Monosomy 10pter, genetic damage of
Monosomy 10q, genetic damage of
Monosomy 11 p11 p12, genetic damage of
Monosomy 11q partial, genetic damage of
Monosomy 12p12 p11, genetic damage of
Monosomy 12p13, genetic damage of
Monosomy 13q, genetic damage of
Monosomy 13q14, genetic damage of
Monosomy 13q22, genetic damage of
Monosomy 13q22, genetic damage of
Monosomy 13q32, genetic damage of
Monosomy 14q11, genetic damage of
Monosomy 14q31, genetic damage of
Monosomy 14qter, genetic damage of
Monosomy 15q1, genetic damage of
Monosomy 15q25, genetic damage of
Monosomy 17q23 q24, genetic damage of
Monosomy 18 mosaicism, genetic damage of
Monosomy 18p, genetic damage of
Monosomy 18q, genetic damage of

Monosomy 18q23, genetic damage of
Monosomy 1p, genetic damage of
Monosomy 1p22 p13, genetic damage of
Monosomy 1p31 p22, genetic damage of
Monosomy 1p32, genetic damage of
Monosomy 1p34 p32, genetic damage of
Monosomy 1q21 q25, genetic damage of
Monosomy 1q25 q32, genetic damage of
Monosomy 1q32 q42, genetic damage of
Monosomy 1q4, genetic damage of
Monosomy 20p, genetic damage of
Monosomy 21, genetic damage of
Monosomy 21q22, genetic damage of
Monosomy 2p22, genetic damage of
Monosomy 2pter p24, genetic damage of
Monosomy 2q, genetic damage of
Monosomy 2q duplication 1p, genetic damage of
Monosomy 2q24, genetic damage of
Monosomy 2q37, genetic damage of
Monosomy 3p, genetic damage of
Monosomy 3p14 p11, genetic damage of
Monosomy 3p2, genetic damage of
Monosomy 3p25, genetic damage of
Monosomy 3q13, genetic damage of
Monosomy 3q21 23, genetic damage of
Monosomy 3q27, genetic damage of
Monosomy 4p, genetic damage of
Monosomy 4p14 p16, genetic damage of
Monosomy 4q, genetic damage of
Monosomy 4q32, genetic damage of
Monosomy 5q35, genetic damage of
Monosomy 6p23, genetic damage of
Monosomy 6q, genetic damage of

Monosomy 6q1, genetic damage of
Monosomy 6q13 q15, genetic damage of
Monosomy 6q16 q21, genetic damage of
Monosomy 6q2, genetic damage of
Monosomy 7, genetic damage of
Monosomy 7q21, genetic damage of
Monosomy 7q3, genetic damage of
Monosomy 8p, genetic damage of
Monosomy 8p23 1, genetic damage of
Monosomy 8q, genetic damage of
Monosomy 8q12 21, genetic damage of
Monosomy 8q21 q22, genetic damage of
Monosomy 9p, genetic damage of
Monosomy X, genetic damage of
Monosomy Xp22 pter, genetic damage of
Monosomy Xq28, genetic damage of
Montefiore syndrome, genetic damage of
Moore Federman syndrome, genetic damage of
Moore Smith Weaver syndrome, genetic damage of
Moore Weaver syndrome, genetic damage of
Morel's ear, genetic damage of
Moreno Zachai Kaufman syndrome, genetic damage of
Morgani Turner Albright syndrome, genetic damage of
Morhosseini Holmes Walton syndrome, genetic damage of
Morillo Cucci Passarge syndrome, genetic damage of
Morphea scleroderma, genetic damage of
Morphea, generalized, genetic damage of
Morquio disease, type A, genetic damage of
Morquio disease, type B, genetic damage of
Morquio syndrome, genetic damage of
Morrison Young syndrome, genetic damage of
Morse Rawnsley Sargent syndrome, genetic damage of
Mosaic trisomy 16, genetic damage of

Mosaic variegated aneuploidy microcephaly syndrome, genetic damage of
Motor neuro-ophthalmic disorders, genetic damage of
Motor neuron disease, genetic damage of
Motor neuropathy, genetic damage of
Motor neuropathy peripheral dysautonomia, genetic damage of
Motor sensory neuropathy type 1 aplasia cutis congenita, genetic damage of
Motorphobia, genetic damage of (possible)
Mounier-Kuhn syndrome, genetic damage of
Mount Reback syndrome, genetic damage of
Mousa Al din Al Nassar syndrome, genetic damage of
Moyamoya disease, genetic damage of
Moynahan syndrome, genetic damage of
MPO deficiency, genetic damage of
MPS 1-H, genetic damage of
MPS 1-H/S, genetic damage of
MPS 1-S, genetic damage of
MPS II (mild), genetic damage of
MPS II (severe), genetic damage of
MPS III-A, genetic damage of
MPS III-B, genetic damage of
MPS III-C, genetic damage of
MPS III-D, genetic damage of
MPS IV-A, genetic damage of
MPS IV-B, genetic damage of
MPS VI, genetic damage of
MPS VII, genetic damage of
MR, genetic damage of
MRKH syndrome, genetic damage of
MSBD syndrome, genetic damage of
MTHFR deficiency, genetic damage of
Mucha-Habermann disease, genetic damage of

Muckle-Wells syndrome, genetic damage of
Mucocutaneous lymph node syndrome, genetic damage of
Mucoepithelial dysplasia, genetic damage of
Mucolipidosis type 1, genetic damage of
Mucolipidosis type 3, genetic damage of
Mucolipidosis type 4, genetic damage of
Mucopolysaccharidosis, genetic damage of
Mucopolysaccharidosis type 3, genetic damage of
Mucopolysaccharidosis type 4, genetic damage of
Mucopolysaccharidosis type I Hurler syndrome, genetic damage of
Mucopolysaccharidosis type I Hurler/Scheie syndrome, genetic damage of
Mucopolysaccharidosis type I Scheie syndrome, genetic damage of
Mucopolysaccharidosis type II Hunter syndrome, genetic damage of
Mucopolysaccharidosis type III-A Sanfilippo syndrome, genetic damage of
Mucopolysaccharidosis type III-B, genetic damage of
Mucopolysaccharidosis type III-C, genetic damage of
Mucopolysaccharidosis type III-D, genetic damage of
Mucopolysaccharidosis type IV-A Morquio syndrome, genetic damage of
Mucopolysaccharidosis type IV-B, genetic damage of
Mucopolysaccharidosis type V, genetic damage of
Mucopolysaccharidosis type VI Maroteaux-Lamy—severe, intermediate, genetic damage of
Mucopolysaccharidosis type VII Sly syndrome, genetic damage of
Mucosulfatidosis, genetic damage of
Muenke syndrome, genetic damage of
Muir-Torre syndrome, genetic damage of
Mulibrey Nanism syndrome, genetic damage of
Muller Barth Menger syndrome, genetic damage of
Mullerian agenesis, genetic damage of

Mullerian aplasia, genetic damage of
Mullerian derivatives lymphangiectasia polydactyly, genetic damage of
Mullerian derivatives, persistent, genetic damage of
Mullerian duct abnormalities galactosemia, genetic damage of
Mullerian duct failure, genetic damage of
Mulliez Roux Loterman syndrome, genetic damage of
Multi-infarct dementia, genetic damage of
Multicentric osteolysis nephropathy, genetic damage of
Multicentric reticulohistiocytosis, genetic damage of
Multifocal heterotopia, genetic damage of
Multifocal motor neuropathy with conduction block, genetic damage of
Multifocal ventricular premature beats, genetic damage of
Multinodular goiter cystic kidney polydactyly, genetic damage of
Multiple acyl-CoA deficiency, genetic damage of
Multiple carboxylase deficiency, biotin responsive, genetic damage of
Multiple carboxylase deficiency, late onset, genetic damage of
Multiple carboxylase deficiency, propionic acidemia, genetic damage of
Multiple chemical sensitivity, genetic damage of
Multiple congenital anomalies mental retardation, growth failure and cleft lip palate, genetic damage of
Multiple congenital contractures, genetic damage of
Multiple contracture syndrome Finnish type, genetic damage of
Multiple endocrine neoplasia type 1, genetic damage of
Multiple endocrine neoplasia, type 2, genetic damage of
Multiple fibrofolliculoma familial, genetic damage of
Multiple hamartoma syndrome, genetic damage of
Multiple hereditary exostoses, genetic damage of
Multiple joint dislocations metaphyseal dysplasia, genetic damage of

Multiple myeloma, genetic damage of
Multiple organ failure, genetic damage of
Multiple pterygium syndrome, genetic damage of
Multiple pterygium syndrome lethal type, genetic damage of
Multiple sclerosis, genetic damage of
Multiple sclerosis ichthyosis factor VIII deficiency, genetic damage of
Multiple subcutaneous angiolipomas, genetic damage of
Multiple sulfatase deficiency, genetic damage of
Multiple synostosis syndrome, genetic damage of
Multiple system atrophy, genetic damage of
Multiple vertebral anomalies unusual facies, genetic damage of
Mulvihill Smith syndrome, genetic damage of
Munchausen by proxy syndrome, genetic damage of
MURCS association, genetic damage of
Muscle-eye-brain syndrome, genetic damage of
Muscular atrophy ataxia retinitis pigmentosa diabetes mellitus, genetic damage of
Muscular dystrophy, genetic damage of
Muscular dystrophy congenital infantile cataract hypogonadism, genetic damage of
Muscular dystrophy congenital, merosin negative, genetic damage of
Muscular dystrophy facioscapulohumeral, genetic damage of
Muscular dystrophy Hutterite type, genetic damage of
Muscular dystrophy limb girdle type 2A, Erb type, genetic damage of
Muscular dystrophy limb-girdle autosomal dominant, genetic damage of
Muscular dystrophy limb-girdle type 2B, Myoshi type, genetic damage of
Muscular dystrophy limb-girdle with beta-sarcoglycan deficiency, genetic damage of

Muscular dystrophy limb-girdle with delta-sarcoglyan deficiency, genetic damage of

Muscular dystrophy limb-girdle with gamma-sarcoglycan deficiency, genetic damage of

Muscular dystrophy white matter spongiosis, genetic damage of

Muscular dystrophy, congenital, merosin-positive, genetic damage of

Muscular dystrophy, Duchenne and Becker type, genetic damage of

Muscular fibrosis multifocal obstructed vessels, genetic damage of

Muscular phosphorylase kinase deficiency, genetic damage of

Mutations in estradiol receptor, genetic damage of

Myalgia eosinophilia associated with tryptophan, genetic damage of

Myalgic encephalomyelitis, genetic damage of

Myasthenia, familial, genetic damage of

Mycophobia, genetic damage of (possible)

Mycosis fungoides, familial, genetic damage of

Myelinopathies, genetic damage of

Myelitis, genetic damage of

Myelocerebellar disorder, genetic damage of

Myelodysplasia, genetic damage of

Myelodysplastic syndromes, genetic damage of

Myelofibrosis, genetic damage of

Myelofibrosis, idiopathic, genetic damage of

Myelofibrosis-osteosclerosis, genetic damage of

Myeloid splenomegaly, genetic damage of

Myeloperoxidase deficiency, genetic damage of

Myhre Ruvalcaba Graham syndrome, genetic damage of

Myhre Ruvalcaba Kelley syndrome, genetic damage of

Myhre School syndrome, genetic damage of

Myhre syndrome, genetic damage of

Myoadenylate deaminase deficiency, genetic damage of

Myocardium disorder, genetic damage of
Myoclonic dystonia, genetic damage of
Myoclonic progressive familial epileps, genetic damage of
Myoclonus, genetic damage of
Myoclonus ataxia, genetic damage of
Myoclonus cerebellar ataxia deafness, genetic damage of
Myoclonus epilepsy, genetic damage of
Myoclonus epilepsy partial seizure, genetic damage of
Myoclonus hereditary progressive distal muscular atrophy, genetic damage of
Myoclonus progressive epilepsy of Unverricht and Lundborg, genetic damage of
Myoclonus with epilepsy with ragged red fibers (mitochondria), genetic damage of
Myofibrillar lysis, genetic damage of
Myofibroblastic tumors, genetic damage of
Myoglobinuria, genetic damage of
Myoglobinuria dominant form, genetic damage of
Myoglobinuria recurrent, genetic damage of
Myoneurogastrointestinal encephalopathy syndrome, genetic damage of
Myopathy, genetic damage of
Myopathy and diabetes mellitus, genetic damage of
Myopathy cataract hypogonadism, genetic damage of
Myopathy congenital multicore with external ophthalmoplegia, genetic damage of
Myopathy growth and mental retardation hypospadias, genetic damage of
Myopathy Hutterite type, genetic damage of
Myopathy mitochondrial cataract, genetic damage of
Myopathy Moebius Robin syndrome, genetic damage of
Myopathy ophthalmoplegia hypoacousia areflexia, genetic damage of

Myopathy tubular agregates, genetic damage of
Myopathy with lactic acidosis and sideroblastic anemia, genetic damage of
Myopathy with lysis of myofibrils, genetic damage of
Myopathy, desmin storage, genetic damage of
Myopathy, McArdle type, genetic damage of
Myopathy, myotubular, genetic damage of
Myopathy, X linked, with excessive autophagy, genetic damage of
Myophosphorylase deficiency, genetic damage of
Myopia, infantile severe, genetic damage of
Myopia, severe, genetic damage of
Myositis, genetic damage of
Myositis ossificans, genetic damage of
Myositis ossificans post-traumatic, genetic damage of
Myositis ossificans progressiva, genetic damage of
Myositis, inclusion body, genetic damage of
Myotonia atrophica, genetic damage of
Myotonia congenita, genetic damage of
Myotonia mental retardation skeletal anomalies, genetic damage of
Myotonic dystrophy, genetic damage of
Myxedema, genetic damage of
Myxoid liposarcoma, genetic damage of
Myxoma-spotty pigmentation-endocrine overactivity, genetic damage of

N

N acetyltransferase deficiency, genetic damage of
N syndrome, genetic damage of
N-acetyl glutamate synthetase deficiency, genetic damage of
N-acetyl-alpha-D-galactosaminidase, genetic damage of
N-acetyl-alpha-glucosaminidase sulfamidase deficiency, genetic damage of
N-acetyl-glucosamine 1-phosphotransferase deficiency, genetic damage of
N-acetyl-glucosamine-6-sulfate sulfatase deficiency, genetic damage of
NADH CoQ reductase, deficiency of, genetic damage of
NADH cytochrome B5 reductase deficiency, genetic damage of
NADH diaphorase deficiency, genetic damage of
NADH methemoglobin reductase deficiency, genetic damage of
Naegelli syndrome, genetic damage of
Nager syndrome, genetic damage of
Naguib Richieri Costa syndrome, genetic damage of
Naguib syndrome, genetic damage of
Nail-patella syndrome, genetic damage of
Nakajo Nishimura syndrome, genetic damage of
Nakajo syndrome, genetic damage of
Nakamura Osame syndrome, genetic damage of
NAME syndrome, genetic damage of
Nance-Horan syndrome, genetic damage of
Nanism due to growth hormone combined deficiency, genetic damage of
Nanism due to growth hormone isolated deficiency, genetic damage of
Nanism due to growth hormone isolated deficiency with X linked hypogammaglobulinemia, genetic damage of
Nanism due to growth hormone resistance, genetic damage of
Narcolepsy, genetic damage of

Narcolepsy-cataplexy, genetic damage of

Narrow oral fissure short stature cone shaped epip, genetic damage of

Nasodigitoacoustic syndrome, genetic damage of

Nasopalpebral lipoma coloboma syndrome, genetic damage of

Nasopharyngeal carcinoma, genetic damage of

Nasopharyngeal neoplasm, genetic damage of

Nasopharyngeal teratoma Dandy Walker diaphragmatic hernia, genetic damage of

Natal teeth intestinal pseudoobstruction patent ductus, genetic damage of

Nathalie syndrome, genetic damage of

NBCC, genetic damage of

NBS, genetic damage of

Necrophobia, genetic damage of (possible)

Necrotizing encephalopathy, infantile subacute, genetic damage of

Negative rheumatoid factor polyarthritis, genetic damage of

Nelson syndrome, genetic damage of

Nemaline myopathy, genetic damage of

Neonatal hemochromatosis, genetic damage of

Neonatal ovarian cyst, genetic damage of

Neonatal transient jaundice, genetic damage of

Neopharmaphobia, genetic damage of (possible)

Neophobia, genetic damage of (possible)

Nephophobia, genetic damage of (possible)

Nephritis, IgA type, genetic damage of

Nephroblastoma, genetic damage of

Nephroblastomatosis, fetal ascites,macrosomia and Wilm's tumor, genetic damage of

Nephrocalcinosis, genetic damage of

Nephrogenic diabetes insipidus, genetic damage of

Nephrolithiasis type 2, genetic damage of

Nephronophtisis, genetic damage of

Nephronophtisis familial adult spastic quadriparesis, genetic damage of

Nephropathy deafness hyperparathyroidism, genetic damage of

Nephropathy familial with gout, genetic damage of

Nephropathy familial with hyperuricemia, genetic damage of

Nephrosclerosis, genetic damage of

Nephrosis deafness urinary tract digital malformation, genetic damage of

Nephrosis neuronal dysmigration syndrome, genetic damage of

Nephrotic syndrome ocular anomalies, genetic damage of

Nephrotic syndrome, idiopathic steroid-resistant, genetic damage of

Nerve sheath neoplasm, genetic damage of

Nesidioblastosis of pancreas, genetic damage of

Netherton syndrome ichthyosis, genetic damage of

Neu Laxova syndrome, genetic damage of

Neuhauser Daly Magnelli syndrome, genetic damage of

Neuhauser Eichner Opitz syndrome, genetic damage of

Neural crest tumor, genetic damage of

Neural tube defects X linked, genetic damage of

Neuraminidase beta-galactosidase deficiency, genetic damage of

Neuraminidase deficiency, genetic damage of

Neurasthenia, genetic damage of (possible)

Neurilemmomatosis, genetic damage of

Neuritic plaque, genetic damage of

Neuritis with brachial predilection, genetic damage of

Neuroacanthocytosis, genetic damage of

Neuroaxonal dystrophy renal tubular acidosis, genetic damage of

Neuroaxonal dystrophy, late infantile, genetic damage of

Neuroblastoma, genetic damage of

Neurocutaneous melanosis, genetic damage of

Neuroectodermal endocrine syndrome, genetic damage of

Neuroectodermal tumor, primitive, genetic damage of

Neuroectodermal tumors primitive, genetic damage of
Neuroendocrine cancer, genetic damage of
Neuroendocrine carcinoma of the cervix, genetic damage of
Neuroendocrine tumor, genetic damage of
Neuroepithelioma, genetic damage of
Neurofaciodigitorenal syndrome, genetic damage of
Neurofibrillary tangles, genetic damage of
Neurofibroma, genetic damage of
Neurofibromatosis, genetic damage of
Neurofibromatosis type 1, genetic damage of
Neurofibromatosis type 2, genetic damage of
Neurofibromatosis type 3, genetic damage of
Neurofibromatosis type 6, genetic damage of
Neurofibromatosis-Noonan syndrome, genetic damage of
Neurofibrosarcoma, genetic damage of
Neurogenic hypertension, genetic damage of
Neuroleptic malignant syndrome, genetic damage of
Neuroma biliary tract, genetic damage of
Neuronal heterotopia, genetic damage of
Neuronal interstitial dysplasia, genetic damage of
Neuronal intestinal pseudoobstruction, genetic damage of
Neuronal intranuclear hyaline inclusion disease, genetic damage of
Neuronal intranuclear inclusion disease, genetic damage of
Neuropathy ataxia and retinis pigmentosa, genetic damage of
Neuropathy congenital sensory neurotrophic keratitis, genetic damage of
Neuropathy hereditary motor and sensory lom type, genetic damage of
Neuropathy hereditary with liability to pressure palsies, genetic damage of
Neuropathy motor sensory type 2 deafness mental retardation, genetic damage of
Neuropathy sensory spastic paraplegia, genetic damage of

Neuropathy, hereditary sensory, type I, genetic damage of
Neuropathy, hereditary sensory, type II, genetic damage of
Neurotoxicity syndromes, genetic damage of
Neutral lipid storage myopathy, genetic damage of
Neutropenia and hyperlymphocytosis with large granular lympho-
cytes, genetic damage of
Neutropenia intermittent, genetic damage of
Neutropenia monocytopenia deafness, genetic damage of
Neutropenia, severe chronic, genetic damage of
Nevi flammei, familial multiple, genetic damage of
Nevo syndrome, genetic damage of
Nevus of ota retinitis pigmentosa, genetic damage of
Nevus sebaceus of Jadassohn, genetic damage of
Nezelof's syndrome, genetic damage of
NF-3B, genetic damage of
Nicolaides Baraitser syndrome, genetic damage of
Niemann-Pick, genetic damage of
Niemann-Pick C1 disease, genetic damage of
Niemann-Pick C2 disease, genetic damage of
Niemann-Pick disease type A, genetic damage of
Niemann-Pick Disease Type B, genetic damage of
Niemann-Pick type C, genetic damage of
Niemann-Pick type D, genetic damage of
Night blindness skeletal anomalies unusual facies, genetic damage
of
Night blindness, congenital stationary, genetic damage of
Niikawa Kuroki syndrome, genetic damage of
Nivelon Nivelon Mabille syndrome, genetic damage of
Noble Bass Sherman syndrome, genetic damage of
Noctiphobia, genetic damage of (possible)
Nodular erythema digital changes, genetic damage of
Nomatophobia, genetic damage of (possible)
Non-functioning pancreatic endocrine tumor, genetic damage of

Non-Hodgkin lymphoma, genetic damage of
Non-lissencephalic cortical dysplasia, genetic damage of
Non-small cell cancer, genetic damage of
Nonallergic atopic dermatitis, genetic damage of
Noninsulin-dependent diabetes mellitus with deafness, genetic damage of
Nonketotic hyperglycinemia, genetic damage of
Nonne-Milroy disease, genetic damage of
Nonsmall cell cancer, genetic damage of
Nonsyndromal microcephaly, genetic damage of
Nonsyndromic hereditary hearing impairment, genetic damage of
Noonan like contracture myopathy hyperpyrexia, genetic damage of
Noonan like syndrome, genetic damage of
Noonan syndrome, genetic damage of
Noonanovej, genetic damage of
Norman Roberts lissencephaly syndrome, genetic damage of
Normokaliemic periodic paralysis, genetic damage of
Norrie disease, genetic damage of
Northern epilepsy, genetic damage of
Norum disease, genetic damage of
Nose polyposis, familial, genetic damage of
Nosocomephobia, genetic damage of (possible)
Nosophobia, genetic damage of (possible)
Notalgia paresthetica, genetic damage of
Nova syndrome, genetic damage of
Novak syndrome, genetic damage of
NTD, genetic damage of
Nyctophobia, genetic damage of (possible)
Nyhan syndrome, genetic damage of

O

O Doherty syndrome, genetic damage of
O Donnell Pappas syndrome, genetic damage of
Obesity, genetic damage of
Obesophobia, genetic damage of (possible)
Obsessive-compulsive disorder, genetic damage of
Obstructive asymmetric septal hypertrophy, genetic damage of
Occipital horn syndrome, genetic damage of
Occult spinal dysraphism, genetic damage of
OCD, genetic damage of
Ochoa syndrome, genetic damage of
Ochronosis, hereditary, genetic damage of
Ocular albinism, genetic damage of
Ocular coloboma-imperforate anus, genetic damage of
Ocular convergence spasm, genetic damage of
Ocular melanoma, genetic damage of
Ocular motility disorders, genetic damage of
Ocular toxoplasmosis, genetic damage of
Oculo cerebral dysplasia, genetic damage of
Oculo cerebro acral syndrome, genetic damage of
Oculo cerebro osseous syndrome, genetic damage of
Oculo dento digital dysplasia, genetic damage of
Oculo digital syndrome, genetic damage of
Oculo facio cardio dental syndrome, genetic damage of
Oculo skeletal renal syndrome, genetic damage of
Oculo tricho anal syndrome, genetic damage of
Oculo tricho dysplasia, genetic damage of
Oculo-dento-digital syndrome, genetic damage of
Oculo-gastrointestinal muscular dystrophy, genetic damage of
Oculoauriculofrontonasal syndrome, genetic damage of
Oculoauriculovertebral dysplasia, genetic damage of
Oculocerebral hypopigmentation syndrome Cross type, genetic damage of

Oculocerebral hypopigmentation syndrome type Preus, genetic damage of

Oculocerebral syndrome with hypopigmentation, genetic damage of

Oculocerebrocutaneous syndrome, genetic damage of

Oculocerebrorenal syndrome, genetic damage of

Oculocerebrorenal syndrome of Lowe, genetic damage of

Oculocutaneous albinism, genetic damage of

Oculocutaneous albinism immunodeficiency, genetic damage of

Oculocutaneous albinism type 1, genetic damage of

Oculocutaneous albinism type 2, genetic damage of

Oculocutaneous albinism type 3, genetic damage of

Oculocutaneous albinism, tyrosinase negative, genetic damage of

Oculocutaneous albinism, tyrosinase positive, genetic damage of

Oculocutaneous tyrosinemia, genetic damage of

Oculodental syndrome Rutherfurd syndrome, genetic damage of

Oculodentodigital dysplasia dominant, genetic damage of

Oculodentodigital syndrome, genetic damage of

Oculodentoosseous dysplasia dominant, genetic damage of

Oculodentoosseous dysplasia recessive, genetic damage of

Oculomaxillofacial dysostosis, genetic damage of

Oculomelic amyoplasia, genetic damage of

Oculopalatocerebral dwarfism, genetic damage of

Oculopalatoskeletal syndrome, genetic damage of

Oculopharnygeal muscular dystrophy, genetic damage of

Oculorenocerebellar syndrome, genetic damage of

Odonto onycho dysplasia with alopecia, genetic damage of

Odontoma, genetic damage of

Odontomicronychial dysplasia, genetic damage of

Odontoonychodermal dysplasia, genetic damage of

Odontophobia, genetic damage of (possible)

Odontotrichomelic hypohidrotic dysplasia, genetic damage of

Odynophobia, genetic damage of (possible)

Oeis complex, genetic damage of
Oerter Friedman Anderson syndrome, genetic damage of
OFD syndrome type 8, genetic damage of
OFD syndrome type Figuera, genetic damage of
Ogilvie's syndrome, genetic damage of
Ohaha syndrome, genetic damage of
Ohdo Madokoro Sonoda syndrome, genetic damage of
Oikophobia, genetic damage of (possible)
Okamuto Satomura syndrome, genetic damage of
Olfactophobia, genetic damage of (possible)
Oligodactyly tetramelic postaxial, genetic damage of
Oligomeganephronic renal hypoplasia, genetic damage of
Oligomeganephrony, genetic damage of
Oligophernia, genetic damage of
Oliver McFarlane syndrome, genetic damage of
Oliver syndrome, genetic damage of
Olivopontocerebellar atrophy, genetic damage of
Olivopontocerebellar atrophy deafness, genetic damage of
Olivopontocerebellar atrophy type 1, genetic damage of
Olivopontocerebellar atrophy type 2, genetic damage of
Olivopontocerebellar atrophy type 3, genetic damage of
Ollier disease, genetic damage of
Olmsted syndrome, genetic damage of
Ombrophobia, genetic damage of (possible)
Omenn syndrome, genetic damage of
Omodysplasia, genetic damage of
Omphalocele cleft palate syndrome lethal, genetic damage of
Omphalocele exstrophy imperforate anus, genetic damage of
Omphalomesenteric cyst, genetic damage of
Onat syndrome, genetic damage of
Ondine syndrome, genetic damage of
Oneirophobia, genetic damage of (possible)
Onychonychia hypoplastic distal phalanges, genetic damage of

Onychotrichodysplasia and neutropenia, genetic damage of
Ophthalmophobia, genetic damage of (possible)
Opitz Mollica Sorge syndrome, genetic damage of
Opitz Reynolds Fitzgerald syndrome, genetic damage of
Opitz syndrome, genetic damage of
Oppositional defiant disorder, genetic damage of
Opsismodysplasia, genetic damage of
Opthalmic icthyosis, genetic damage of
Opthalmo acromelic syndrome, genetic damage of
Opthalmomandibulomelic dysplasia, genetic damage of
Opthalmoplegia ataxia hypoacusis, genetic damage of
Opthalmoplegia mental retardation lingua scrotalis, genetic damage of
Opthalmoplegia myalgia tubular aggregates, genetic damage of
Opthalmoplegia progressive external scoliosis, genetic damage of
Optic atrophy, genetic damage of
Optic atrophy opthalmoplegia ptosis deafness myopia, genetic damage of
Optic atrophy polyneuropathy deafness, genetic damage of
Optic atrophy, autosomal dominant, genetic damage of
Optic atrophy, idiopathic, autosomal recessive, genetic damage of
Optic atrophy, Leber type, genetic damage of
Optic nerve coloboma with renal disease, genetic damage of
Optic nerve disorder, genetic damage of
Optic neuritis, genetic damage of
Optic pathway glioma, genetic damage of
Opticoacoustic nerve atrophy dementia, genetic damage of
Oral facial digital syndrome, genetic damage of
Oral facial digital syndrome type 3, genetic damage of
Oral facial digital syndrome type 4, genetic damage of
Oral facial dyskinesia, genetic damage of
Oral leukoplakia, genetic damage of
Oral lichen planus, genetic damage of

Oral lichenoid lesions, genetic damage of
Oral squamous cell carcinoma, genetic damage of
Oral submucous fibrosis, genetic damage of
Oral-facial cleft, genetic damage of
Oral-facial-digital syndrome, genetic damage of
Oral-pharyngeal disorders, genetic damage of
Organic brain syndrome, genetic damage of
Organic mood syndrome, genetic damage of
Organic personality syndrome, genetic damage of
Ornithine aminotransferase deficiency, genetic damage of
Ornithine carbamoyl phosphate deficiency, genetic damage of
Ornithine carbamoyl transferase deficiency, genetic damage of
Ornithinemia, genetic damage of
Oro acral syndrome, genetic damage of
Orocraniodigital syndrome, genetic damage of
Orofaciodigital syndrome Gabrielli type, genetic damage of
Orofaciodigital syndrome Shashi type, genetic damage of
Orofaciodigital syndrome Thurston type, genetic damage of
Orofaciodigital syndrome type 2, genetic damage of
Orofaciodigital syndrome type1, genetic damage of
Orotic aciduria hereditary, genetic damage of
Orotic aciduria purines-pyrimidines, genetic damage of
Orotidylic decarboxylase deficiency, genetic damage of
Osebold Remondini syndrome, genetic damage of
Osgood-Schlatter disease, genetic damage of
Oslam syndrome, genetic damage of
Osler-Weber-Rendu syndrome, genetic damage of
Osmed syndrome, genetic damage of
Ossicular malformations, familial, genetic damage of
Osteitis deformans, genetic damage of
Osteoarthritis, genetic damage of
Osteoarthropathy of fingers familial, genetic damage of
Osteochondritis, genetic damage of

Osteochondritis deformans, genetic damage of
Osteochondritis deformans, genetic damage of
Osteochondritis deformans juvenile, genetic damage of
Osteochondritis dissecans, genetic damage of
Osteochondrodysplasia thrombocytopenia hydrocephalus, genetic damage of
Osteochondroma, genetic damage of
Osteocraniostenosis, genetic damage of
Osteodysplasia familial Anderson type, genetic damage of
Osteodysplastic dwarfism Corsello type, genetic damage of
Osteoectasia familial, genetic damage of
Osteogenesis imperfecta congenita microcephaly and cataracts, genetic damage of
Osteogenesis imperfecta congenital joint contractures, genetic damage of
Osteogenesis imperfecta retinopathy, genetic damage of
Osteogenic sarcoma, genetic damage of
Osteoglophonic dwarfism, genetic damage of
Osteolysis hereditary multicentric, genetic damage of
Osteolysis syndrome recessive, genetic damage of
Osteomalacia, genetic damage of
Osteomyelitis, genetic damage of
Osteonecrosis, genetic damage of
Osteopathia condensans disseminata with osteopoikilosis, genetic damage of
Osteopathia striata cranial sclerosis, genetic damage of
Osteopathia striata pigmentary dermopathy white forelock, genetic damage of
Osteopetrosis autosomal dominant type 1, genetic damage of
Osteopetrosis lethal, genetic damage of
Osteopetrosis renal tubular acidosis, genetic damage of
Osteopetrosis, (generic term), genetic damage of
Osteopetrosis, malignant, genetic damage of

Osteopetrosis, mild autosomal recessive form, genetic damage of
Osteopoikilosis, genetic damage of
Osteoporosis, genetic damage of
Osteoporosis macrocephaly mental retardation blindness, genetic damage of
Osteoporosis oculocutaneous hypopigmentation syndrome, genetic damage of
Osteoporosis pseudoglioma syndrome, genetic damage of
Osteosarcoma, genetic damage of
Osteosarcoma limb anomalies erythroid macrocytosis, genetic damage of
Osteosclerose type Stanescu, genetic damage of
Osteosclerosis, genetic damage of
Osteosclerosis abnormalities of nervous system and meninges, genetic damage of
Osteosclerosis autosomal dominant Worth type, genetic damage of
Ostertag type amyloidosis, genetic damage of
Ostravik Lindemann Solberg syndrome, genetic damage of
Ota Kawamura Ito syndrome, genetic damage of
Oto palato digital syndrome type I and II, genetic damage of
Oto-Palatal-digital syndrome, genetic damage of
Otodental dysplasia, genetic damage of
Otofaciocervical syndrome, genetic damage of
Otoonychoperoneal syndrome, genetic damage of
Otopalatodigital syndrome type 2, genetic damage of
Otosclerosis, genetic damage of
Otosclerosis, familial, genetic damage of
Otospondylomegaepiphyseal dysplasia; OSMED, genetic damage of
Ouvrier Billson syndrome, genetic damage of
Ovarian cancer, genetic damage of
Ovarian carcinosarcoma, genetic damage of
Ovarian dwarfism, genetic damage of

Ovarian dwarfism as part of Turner syndrome, genetic damage of
Ovarian insufficiency due to FSH resistance, genetic damage of
Ovarian remnant syndrome, genetic damage of
Overfolded helix, genetic damage of
Overgrowth radial ray defect arthrogryposis, genetic damage of
Overgrowth syndrome type Fryer, genetic damage of
Overhydrated hereditary stomatocytosis, genetic damage of
Oxalosis, genetic damage of
Oxoglutaricaciduria, genetic damage of

P

Pachydermoperiostosis, genetic damage of
Pachygyria, genetic damage of
Pachyonychia congenita Jackson Lawler type, genetic damage of
Pacman syndrome, genetic damage of
Paes Whelan Modi syndrome, genetic damage of
Paget disease extramammary, genetic damage of
Paget disease juvenile type, genetic damage of
Paget's disease, genetic damage of
Paget's disease of the bone, genetic damage of
Paget's disease of the breast, genetic damage of
Pagon Bird Detter syndrome, genetic damage of
Pagon Stephan syndrome, genetic damage of
Pai Levkoff syndrome, genetic damage of
Pai syndrome, genetic damage of
Palant cleft palate syndrome, genetic damage of
Palindromic rheumatism, genetic damage of
Pallister Mosaic syndrome, genetic damage of
Pallister-Hall syndrome, genetic damage of
Palmer Pagon syndrome, genetic damage of
Palmitoyl-protein thioesterase deficiency, genetic damage of
Palmoplantar keratoderma, genetic damage of
Palmoplantar porokeratosis of Mantoux, genetic damage of
Palsy cerebral, genetic damage of
Pancreas agenesis, genetic damage of
Pancreatic adenoma, genetic damage of
Pancreatic beta cell agenesis with neonatal diabetes mellitus, genetic damage of
Pancreatic cancer, genetic damage of
Pancreatic carcinoma, familial, genetic damage of
Pancreatic diseases, genetic damage of
Pancreatic islet cell neoplasm, genetic damage of
Pancreatic islet cell tumors, genetic damage of

Pancreatic lipomatosis duodenal stenosis, genetic damage of
Pancreatitis, hereditary, genetic damage of
Pancreatoblastoma, genetic damage of
PANDAS, genetic damage of
Panhypopituitarism, genetic damage of
Panic disorder, genetic damage of
Panmyelophthisis aplastic anemia, genetic damage of
Panniculitis, genetic damage of
Panophobia, genetic damage of (possible)
Panostotic fibrous dysplasia, genetic damage of
Panthophobia, genetic damage of (possible)
Papilledema, genetic damage of
Papillion-Lefevre syndrome, genetic damage of
Papillitis, genetic damage of
Papilloma of choroid plexus, genetic damage of
Papular mucinosis, genetic damage of
Papular urticaria, genetic damage of
Paraganglioma, genetic damage of
Paramyotonia congenita, genetic damage of
Paramyotonia congenita of Von Eulenburg, genetic damage of
Paraneoplastic cerebellar degeneration, genetic damage of
Paraomphalocele, genetic damage of
Paraparesis amyotrophy of hands and feet, genetic damage of
Paraplegia, genetic damage of
Paraplegia-brachydactyly-cone shaped epiphysis, genetic damage of
Paraplegia-mental retardation-hyperkeratosis, genetic damage of
Parapsoriasis, genetic damage of
Parastremmatic dwarfism, genetic damage of
Parathyroid cancer, genetic damage of
Parathyroid neoplasm, genetic damage of
PARC syndrome, genetic damage of
Parenchymatous cortical degeneration of cerebellum, genetic dam-

age of
Paris-Trousseau thrombopenia, genetic damage of
Parkes-Weber syndrome, genetic damage of
Parkinson dementia Steele type, genetic damage of
Parkinson's disease, genetic damage of
Parkinsonism, genetic damage of
Parkinsonism early onset mental retardation, genetic damage of
Paroxysmal cold hemoglobinuria, genetic damage of
Paroxysmal dystonic choreoathetosis, genetic damage of
Paroxysmal nocturnal hemoglobinuria, genetic damage of
Paroxysmal ventricular fibrillation, genetic damage of
Parry-Romberg syndrome, genetic damage of
Pars planitis, genetic damage of
Parsonage Turner syndrome, genetic damage of
Partial atrioventricular canal, genetic damage of
Partial deletion of Y, genetic damage of
Partial gigantism in context of NF, genetic damage of
Partial lissencephaly, genetic damage of
Partial trisomy 8, genetic damage of
Partington Anderson syndrome, genetic damage of
Partington Mulley syndrome, genetic damage of
Parturiphobia, genetic damage of (possible)
Pascuel Castroviejo syndrome, genetic damage of
Pashayan syndrome, genetic damage of
Pat1, genetic damage of
Pat11, genetic damage of
Pat111, genetic damage of
Pat12, genetic damage of
Pat121, genetic damage of
Pat122, genetic damage of
Pat13, genetic damage of
Pat131, genetic damage of
Pat132, genetic damage of

Pat14, genetic damage of
Pat141, genetic damage of
Pat142, genetic damage of
Patau syndrome, genetic damage of
Patel Bixler syndrome, genetic damage of
Patella aplasia coxa vara tarsal synostosis, genetic damage of
Patella hypoplasia mental retardation, genetic damage of
Patent ductus arteriosus, genetic damage of
Patent ductus arteriosus familial, genetic damage of
Pathophobia, genetic damage of (possible)
Patterson Lowry syndrome, genetic damage of
Patterson pseudoleprechaunism syndrome, genetic damage of
Patterson Stevenson syndrome, genetic damage of
Pauciarticular chronic arthritis, genetic damage of
Pavone Fiumara Rizzo syndrome, genetic damage of
Pearson's marrow/pancreas syndrome, genetic damage of
Pediatric T-cell leukemia, genetic damage of
Peeling skin syndrome ichthyosis, genetic damage of
Peho syndrome, genetic damage of
Pelizaeus-Merzbacher brain sclerosis, genetic damage of
Pelizaeus-Merzbacher disease, genetic damage of
Pelizaeus-Merzbacher disease, recessive, acute infantile, genetic damage of
Pelizaeus-Merzbacher leukodystrophy, genetic damage of
Pellagra like syndrome, genetic damage of
Pellagrophobia, genetic damage of (possible)
Pelvic dysplasia arthrogryposis of lower limbs, genetic damage of
Pelvic lipomatosis, genetic damage of
Pelvic shoulder dysplasia, genetic damage of
Pemphigus, genetic damage of
Pemphigus and fogo selvagem, genetic damage of
Pemphigus foliaceus, genetic damage of
Pemphigus vulgaris, genetic damage of

Pemphigus vulgaris, familial, genetic damage of
Pena Shokeir syndrome, genetic damage of
Pendred syndrome, genetic damage of
Penis agenesia, genetic damage of
Penoscrotal transposition, genetic damage of
Penta X syndrome, genetic damage of
Pentalogy of Cantrell, genetic damage of
Pentosuria, genetic damage of
Penttinen-Aula syndrome, genetic damage of
PEPCK 1 deficiency, genetic damage of
PEPCK 2 deficiency, genetic damage of
PEPCK deficiency, mitochondrial, genetic damage of
Peptidic growth factors deficiency, genetic damage of
Periarteritis nodosa, genetic damage of
Pericardial constriction growth failure, genetic damage of
Pericardial defect diaphragmatic hernia, genetic damage of
Pericardium absent mental retardation short stature, genetic damage of
Pericardium congenital anomaly, genetic damage of
Perilymphatic fistula, genetic damage of
Perimyositis, genetic damage of
Peripartum cardiomyopathy, genetic damage of
Peripheral blood vessel disorder, genetic damage of
Peripheral nervous disorder, genetic damage of
Peripheral neuroectodermal tumor, genetic damage of
Peripheral neuropathy, genetic damage of
Peripheral T-cell lymphoma, genetic damage of
Peripheral type neurofibromatosis, genetic damage of
Perisylvian syndrome congenital bilateral, genetic damage of
Periventricular laminar heterotropia, genetic damage of
Periventricular nodular heterotopia, genetic damage of
Pernicious anemia, genetic damage of
Perniola Krajewska Carnevale syndrome, genetic damage of

Perniosis, genetic damage of
Peroxisomal bifunctional enzyme deficiency, genetic damage of
Perrault syndrome, genetic damage of
Persistent Mullerian duct syndrome, genetic damage of
Persistent truncus arteriosus, genetic damage of
Pes planus, genetic damage of
Peters anomaly, genetic damage of
Peters anomaly with short limb dwarfism, genetic damage of
Peters congenital glaucoma, genetic damage of
Peters-plus syndrome, genetic damage of
Petit Fryns syndrome, genetic damage of
Petty Laxova Wiedemann syndrome, genetic damage of
Peutz-Jeghers syndrome, genetic damage of
Peyronie disease, genetic damage of
Pfeiffer cardiocranial syndrome, genetic damage of
Pfeiffer Hirschfelder Rott syndrome, genetic damage of
Pfeiffer Kapferer syndrome, genetic damage of
Pfeiffer Mayer syndrome, genetic damage of
Pfeiffer Palm Teller syndrome, genetic damage of
Pfeiffer Rockelein syndrome, genetic damage of
Pfeiffer Singer Zschiesche syndrome, genetic damage of
Pfeiffer syndrome, genetic damage of
Pfeiffer Tietze Welte syndrome, genetic damage of
Pfeiffer type acrocephalosyndactyly, genetic damage of
Phacomatosis fourth, genetic damage of
Phacomatosis pigmentokeratotica, genetic damage of
Phacomatosis pigmentovascularis, genetic damage of
Phalacrophobia, genetic damage of (possible)
Phaoke Sharma Agarawal syndrome, genetic damage of
Pharmacophobia, genetic damage of (possible)
Phaver syndrome, genetic damage of
Phenylalanine hydroxylase deficiency, genetic damage of
Phenylalaninemia, genetic damage of

Phenylketonuria, genetic damage of
Phenylketonuria type II, genetic damage of
Phenylketonurias, genetic damage of
Phenylketonuric embryopathy, genetic damage of
Pheochromocytoma, genetic damage of
Pheochromocytoma as part of NF, genetic damage of
Philadelphia-negative chronic myeloid leukemia, genetic damage of
Phocomelia contractures absent thumb, genetic damage of
Phocomelia ectrodactyly deafness sinus arrhythmia, genetic damage of
Phocomelia Schinzel type, genetic damage of
Phocomelia syndrome, genetic damage of
Phocomelia thrombocytopenia encephalocele, genetic damage of
Phosphate diabetes, genetic damage of
Phosphoenolpyruvate carboxykinase 1 deficiency, genetic damage of
Phosphoenolpyruvate carboxykinase 2 deficiency, genetic damage of
Phosphoenolpyruvate carboxykinase deficiency, genetic damage of
Phosphoglucomutase deficiency, genetic damage of
Phosphoglucomutase deficiency type 1, genetic damage of
Phosphoglucomutase deficiency type 2, genetic damage of
Phosphoglucomutase deficiency type 3, genetic damage of
Phosphoglucomutase deficiency type 4, genetic damage of
Phosphoglycerate kinase 1 deficiency, genetic damage of
Phosphoglycerate kinase deficiency, genetic damage of
Phosphomannoisomerase deficiency, genetic damage of
Phosphoribosylpyrophosphate synthetase deficiency, genetic damage of
Photoaugliaphobia, genetic damage of (possible)
Photosensitive epilepsy, genetic damage of
Phthiriophobia, genetic damage of (possible)

Physical urticaria, genetic damage of
Phytanic acid oxidase deficiency, genetic damage of
Phytosterolemia, genetic damage of
PIBIDS syndrome, genetic damage of
Pica, genetic damage of
Picardi-Lassueur-Little syndrome, genetic damage of
Pick disease of the brain, genetic damage of
Pie Torcido, genetic damage of
Piebald trait neurologic defects, genetic damage of
Piebaldism, genetic damage of
Piepkorn Karp Hickoc syndrome, genetic damage of
Pierre Marie cerbellar ataxia, genetic damage of
Pierre Robin sequence congenital heart defect talipes, genetic damage of
Pierre Robin sequence faciodigital anomaly, genetic damage of
Pierre Robin syndrome fetal chondrodysplasia, genetic damage of
Pierre Robin syndrome hyperphalangy clinodactyly, genetic damage of
Pierre Robin syndrome oligodactyly, genetic damage of
Pierre Robin syndrome skeletal dysplasia polydactyly, genetic damage of
Pierre Robin's sequence, genetic damage of
Pierre-Robin syndrome, genetic damage of
Pigment-dispersion syndrome, genetic damage of
Pigmentary retinopathy, genetic damage of
Pigmented villonodular synovitis, genetic damage of
Pignata guarino syndrome, genetic damage of
Pili canulati, genetic damage of
Pili multigemini, genetic damage of
Pili torti, genetic damage of
Pili torti developmental delay neurological abnormalities, genetic damage of
Pili torti nerve deafness, genetic damage of

Pili torti onychodysplasia, genetic damage of
Pillay syndrome, genetic damage of
Pilo dento ungular dysplasia microcephaly, genetic damage of
Pilotto syndrome, genetic damage of
Pinealoma, genetic damage of
Pinheiro Freire Maia Miranda syndrome, genetic damage of
Pinsky Di George Harley syndrome, genetic damage of
Pipecolic acidemia, genetic damage of
PIRA, genetic damage of
Pitt Hopkins syndrome, genetic damage of
Pitt-Rogers-Danks syndrome, genetic damage of
Pituitary dwarfism, genetic damage of
Pityriasis lichenoides chronica, genetic damage of
Pityriasis rubra pilaris, genetic damage of
Piussan Lenaerts Mathieu syndrome, genetic damage of
Placenta disorders, genetic damage of
Placenta neoplasm, genetic damage of
Plagiocephaly X linked mental retardation, genetic damage of
Plasmacytoma anaplastic, genetic damage of
Plasmalogenes synthesis deficiency isolated, genetic damage of
Plasminogen activitor inhibitor type 1 deficiency, congenital, genetic damage of
Plasminogen deficiency, congenital, genetic damage of
Platelet disorder, genetic damage of
Platyspondylic lethal chondrodysplasia, genetic damage of
Platyspondyly amelogenesis imperfecta, genetic damage of
Plexosarcoma, genetic damage of
Plott syndrome, genetic damage of
Plum syndrome, genetic damage of
Plummer-Vinson syndrome, genetic damage of
Podder-Tolmie syndrome, genetic damage of
POEMS syndrome, genetic damage of
Poikiloderma atrophicans-cataract, genetic damage of

Poikiloderma congenital with bullae Weary type, genetic damage of
Poikiloderma hereditary acrokeratotic Weary type, genetic damage of
Poikiloderma of Kindler, genetic damage of
Poikiloderma of Rothmund-Thomson, genetic damage of
Poikilodermatomyositis mental retardation, genetic damage of
Poikilodermia alopecia retrognathism cleft palate, genetic damage of
Pointer syndrome, genetic damage of
Poland anomaly, genetic damage of
Poland syndrome, genetic damage of
Poliosophobia, genetic damage of (possible)
Polyarteritis, genetic damage of
Polyarteritis nodosa, genetic damage of
Polyarthritis, genetic damage of
Polyarthritis, systemic, genetic damage of
Polychondritis, genetic damage of
Polycystic kidney disease, genetic damage of
Polycystic kidney disease, adult type, genetic damage of
Polycystic kidney disease, dominant type, genetic damage of
Polycystic kidney disease, infantile type, genetic damage of
Polycystic kidney disease, recessive type, genetic damage of
Polycystic kidney disease, type 1, genetic damage of
Polycystic kidney disease, type 2, genetic damage of
Polycystic kidney disease, type 3, genetic damage of
Polycystic ovarian disease, familial, genetic damage of
Polycystic ovarian syndrome, genetic damage of
Polycystic ovaries urethral sphincter dysfunction, genetic damage of
Polycythemia vera, genetic damage of
Polydactyly, genetic damage of
Polydactyly alopecia seborrheic dermatitis, genetic damage of

Polydactyly cleft lip palate psychomotor retardation, genetic damage of

Polydactyly myopia syndrome, genetic damage of

Polydactyly postaxial, genetic damage of

Polydactyly postaxial dental and vertebral, genetic damage of

Polydactyly postaxial with median cleft of upper lip, genetic damage of

Polydactyly preaxial type 1, genetic damage of

Polydactyly syndrome middle ray duplication, genetic damage of

Polydactyly visceral anomalies cleft lip palate, genetic damage of

Polyglucosan body disease, adult, genetic damage of

Polymicrogyria turricephaly hypogenitalism, genetic damage of

Polymorphic catecholergic ventricular tachycardia, genetic damage of

Polymorphic macular degeneration, genetic damage of

Polymorphous low-grade adenocarcinoma, genetic damage of

Polymyalgia rheumatica, genetic damage of

Polymyositis, genetic damage of

Polyneuritis, genetic damage of

Polyneuropathy hand defect, genetic damage of

Polyneuropathy mental retardation acromicria prema, genetic damage of

Polyostotic fibrous dysplasia, genetic damage of

Polyposis hamartomatous intestinal, genetic damage of

Polyposis skin pigmentation alopecia fingernail changes, genetic damage of

Polyposis, familial, genetic damage of (Polyposis, familial is one of a large number of polyposis syndromes)

Polysyndactyly cardiac malformation, genetic damage of

Polysyndactyly microcephaly ptosis, genetic damage of

Polysyndactyly orofacial anomalies, genetic damage of

Polysyndactyly overgrowth syndrome, genetic damage of

Polysyndactyly trigonocephaly agenesis of corpus callosum, genetic

damage of
Polysyndactyly type 4, genetic damage of
Polysyndactyly type Haas, genetic damage of
Poncet-Spiegler's cylindroma, genetic damage of
Pontoneocerebellar hypoplasia, genetic damage of
Popliteal pterygium syndrome, genetic damage of
Popliteal pterygium syndrome lethal type, genetic damage of
Porencephaly, genetic damage of
Porencephaly cerebellar hypoplasia malformations, genetic damage of
Porokeratosis of Mibelli, genetic damage of
Porokeratosis plantaris palmaris et disseminata, genetic damage of
Porokeratosis punctata palmaris et plantaris, genetic damage of
Porphyria, genetic damage of
Porphyria cutanea tarda, genetic damage of
Porphyria cutanea tarda, familial type, genetic damage of
Porphyria cutanea tarda, sporadic type, genetic damage of
Porphyria, acute intermittent, genetic damage of
Porphyria, Ala-D, genetic damage of
Porphyria, congenital erythropoietic, genetic damage of
Porphyria, hereditary coproporphyria, genetic damage of
Port wine nevi mega cisterna magna hydrocephalus, genetic damage of
Portal hypertension, genetic damage of
Portal hypertension due to infrahepatic block, genetic damage of
Portal thrombosis, genetic damage of
Portal vein thrombosis, genetic damage of
Portuguese type amyloidosis, genetic damage of
Positive rheumatoid factor polyarthritis, genetic damage of
Postaxial polydactyly mental retardation, genetic damage of
Posterior tibial tendon rupture, genetic damage of
Posterior urethral valves, genetic damage of
Posterior uveitis, genetic damage of

Posterior valve urethra, genetic damage of
Postural hypotension, genetic damage of
Potassium aggravated myotonia, genetic damage of
Potophobia, genetic damage of (possible)
Potter disease type 1, genetic damage of
Potter disease, type 3, genetic damage of
Potter sequence cleft cardiopathy, genetic damage of
Potter syndrome, genetic damage of
Potter syndrome dominant type, genetic damage of
Powell Buist Stenzel syndrome, genetic damage of
Powell Chandra Saal syndrome, genetic damage of
Powell Venencie Gordon syndrome, genetic damage of
Prader-Willi syndrome, genetic damage of
Prata Liberal Goncalves syndrome, genetic damage of
Preaxial deficiency postaxial polydactyly hypospadia, genetic damage of
Preaxial polydactyly colobomata mental retardation, genetic damage of
Precocious epileptic encephalopathy, genetic damage of
Precocious myoclonic encephalopathy, genetic damage of
Precocious puberty, genetic damage of
Precocious puberty, gonadotropin-dependant, genetic damage of
Precocious puberty, male limited, genetic damage of
Preeclampsia, genetic damage of
Preeyasombat Viravithya syndrome, genetic damage of
Pregnancy toxemia /hypertension, genetic damage of
Prekallikrein deficiency, congenital, genetic damage of
Premature aging, genetic damage of
Premature aging, Okamoto type, genetic damage of
Premature atherosclerosis photomyoclonic epilepsy, genetic damage of
Premature menopause, familial, genetic damage of
Premature ovarian failure, genetic damage of

Presbycusis, genetic damage of
Prieto Badia Mulas syndrome, genetic damage of
Prieur Griscelli syndrome, genetic damage of
Primary agammaglobulinemia, genetic damage of
Primary aldosteronism, genetic damage of
Primary amenorrhea, genetic damage of
Primary biliary cirrhosis, genetic damage of
Primary ciliary dyskinesia, genetic damage of
Primary craniosynostosis, genetic damage of
Primary cutaneous amyloidosis, genetic damage of
Primary granulocytic sarcoma, genetic damage of
Primary hyperoxaluria, genetic damage of
Primary lateral sclerosis, genetic damage of
Primary malignant lymphoma, genetic damage of
Primary pulmonary hypertension, genetic damage of
Primary sclerosing cholangitis, genetic damage of
Primary tubular proximal acidosis, genetic damage of
Primerose syndrome, genetic damage of
Primordial dwarfism, genetic damage of
Primordial microcephalic dwarfism Crachami type, genetic damage of
Prinzmetal's variant angina, genetic damage of
Procarcinoma, genetic damage of
Proconvertin deficiency, congenital, genetic damage of
Progeria, genetic damage of
Progeria short stature pigmented nevi, genetic damage of
Progeria variant syndrome Ruvalcaba type, genetic damage of
Progeroid syndrome De Barsy type, genetic damage of
Progeroid syndrome Petty type, genetic damage of
Progeroid syndrome, Penttinen type, genetic damage of
Prognathism dominant, genetic damage of
Progressive acromelanosis, genetic damage of
Progressive black carbon hyperpigmentation of infancy, genetic

damage of
Progressive diaphyseal dysplasia, genetic damage of
Progressive external ophthalmoplegia, genetic damage of
Progressive hearing loss stapes fixation, genetic damage of
Progressive kinking of the hair, genetic damage of
Progressive multifocal leukoencephalopathy, genetic damage of
Progressive myositis ossificans, genetic damage of
Progressive osseous heteroplasia, genetic damage of
Progressive spinal muscular atrophy, genetic damage of
Progressive supranuclear palsy, genetic damage of
Progressive supranuclear palsy atypical, genetic damage of
Progressive systemic sclerosis, genetic damage of
Prolactinoma, familial, genetic damage of
Prolerating trichilemmal cyst, genetic damage of
Prolidase deficiency, genetic damage of
Proline oxidase deficiency, genetic damage of
Prolymphocytic leukemia, genetic damage of
Properdin deficiency, genetic damage of
Propionic acidemia, genetic damage of
Propionyl-CoA carboxylase deficiency, genetic damage of
Prosencephaly cerebellar dysgenesis, genetic damage of
Prostaglandin antenatal infection, genetic damage of
Prostate cancer, familial, genetic damage of
Prostatic malacoplakia associated with prostatic abscess, genetic
damage of
Prostatitis, genetic damage of
Protein C deficiency, genetic damage of
Protein R deficiency, genetic damage of
Protein S deficiency, genetic damage of
Proteus like syndrome mental retardation eye defect, genetic dam-
age of
Proteus syndrome, genetic damage of
Prothrombin deficiency, genetic damage of

Protoporphyria, genetic damage of
Protoporphyria, erythropoietic, genetic damage of
Proud Levine Carpenter syndrome, genetic damage of
Proximal myotonic dystrophy, genetic damage of
Proximal myotonic myopathy, genetic damage of
Proximal spinal muscular atrophy, genetic damage of
Proximal tubulopathy diabetes mellitus cerebellar ataxia, genetic damage of
Prune belly syndrome, genetic damage of
Prurigo nodularis, genetic damage of
Psellismophobia, genetic damage of (possible)
Pseudo-Gaucher disease, genetic damage of
Pseudo-Pelade of Brocq, genetic damage of
Pseudo-Turner syndrome, genetic damage of
Pseudo-Zellweger syndrome, genetic damage of
Pseudoachondroplasia, genetic damage of
Pseudoachondroplastic dysplasia, genetic damage of
Pseudoachondroplastic dysplasia 1, genetic damage of
Pseudoadrenoleukodystrophy, genetic damage of
Pseudoaminopterin syndrome, genetic damage of
Pseudoarylsulfatase A deficiency, genetic damage of
Pseudocholinesterase deficiency, genetic damage of
Pseudogout, genetic damage of
Pseudohermaphrodism anorectal anomalies, genetic damage of
Pseudohermaphroditism, genetic damage of
Pseudohermaphroditism female skeletal anomalies, genetic damage of
Pseudohermaphroditism male with gynecomastia, genetic damage of
Pseudohermaphroditism mental retardation, genetic damage of
Pseudohypoaldosteronism, genetic damage of
Pseudohypoaldosteronism type 1, genetic damage of
Pseudohypoaldosteronism type 2, genetic damage of

Pseudohypoparathyroidism, genetic damage of
Pseudomarfanism, genetic damage of
Pseudomongolism, genetic damage of
Pseudomyxoma peritonei, genetic damage of
Pseudoobstruction idiopathic intestinal, genetic damage of
Pseudopapilledema blepharophimosis hand anomalies, genetic damage of
Pseudoprogeria syndrome, genetic damage of
Pseudotoxoplasmosis syndrome, genetic damage of
Pseudotumor cerebri, genetic damage of
Pseudovaginal perineoscrotal hypospadias, genetic damage of
Pseudoxanthoma elasticum, genetic damage of
Pseudoxanthoma elasticum, dominant form, genetic damage of
Pseudoxanthoma elasticum, recessive form, genetic damage of
Psoriasis, genetic damage of
Psoriatic arthritis, genetic damage of
Psoriatic rheumatism, genetic damage of
Psychophysiologic disorders, genetic damage of
Pterigium Colli, genetic damage of
Pteromerhanophobia, genetic damage of (possible)
Pterygia mental retardation facial dysmorphism, genetic damage of
Pterygium colli mental retardation digital anomalies, genetic damage of
Pterygium of the conjunctiva, genetic damage of
Pterygium syndrome antecubital, genetic damage of
Pterygium syndrome multiple dominant type, genetic damage of
Pterygium syndrome X linked, genetic damage of
Pterygium syndrome, multiple, genetic damage of
Ptosis coloboma mental retardation, genetic damage of
Ptosis coloboma trigonocephaly, genetic damage of
Ptosis strabismus diastasis, genetic damage of
Ptosis strabismus ectopic pupils, genetic damage of
Pulmonar arterioveinous aneurysm, genetic damage of

Pulmonary agenesis, genetic damage of
Pulmonary alveolar proteinosis, genetic damage of
Pulmonary alveolar proteinosis, congenital, genetic damage of
Pulmonary arterio-veinous fistula, genetic damage of
Pulmonary artery agenesis, genetic damage of
Pulmonary artery coming from the aorta, genetic damage of
Pulmonary artery familial dilatation, genetic damage of
Pulmonary atresia with ventricular septal defect, genetic damage of
Pulmonary blastoma, genetic damage of
Pulmonary branches stenosis, genetic damage of
Pulmonary cystic lymphangiectasis, genetic damage of
Pulmonary fibrosis /granuloma, genetic damage of
Pulmonary hypertension, genetic damage of
Pulmonary hypertension, secondary, genetic damage of
Pulmonary hypoplasia familial primary, genetic damage of
Pulmonary sequestration, genetic damage of
Pulmonary supravalvular stenosis, genetic damage of
Pulmonary surfactant protein B, deficiency of, genetic damage of
Pulmonary valve stenosis, genetic damage of
Pulmonary valves agenesis, genetic damage of
Pulmonary veins stenosis, genetic damage of
Pulmonary veno-occlusive disease, genetic damage of
Pulmonary venous return anomaly, genetic damage of
Pulmonaryatresia intact ventricular septum, genetic damage of
Pulmonic stenosis with cafe-au-lait spots, genetic damage of
Punctate acrokeratoderma freckle like pigmentation, genetic damage of
Punctate inner choroidopathy, genetic damage of
Pupaphobia, genetic damage of (possible)
Pure red cell aplasia, genetic damage of
Puretic syndrome, genetic damage of
Purine nucleoside phosphorylase deficiency, genetic damage of
Purpura, genetic damage of

Purpura, Schoenlein-Henoch, genetic damage of
Purpura, thrombotic thrombocytopenic, genetic damage of
Purtilo syndrome, genetic damage of
Pycnodysostosis, genetic damage of
Pyknoachondrogenesis, genetic damage of
Pyle disease, genetic damage of
Pyoderma gangrenosum, genetic damage of
Pyrexiophobia, genetic damage of (possible)
Pyridoxine deficit, genetic damage of
Pyrimidinemia familial, genetic damage of
Pyrophobia, genetic damage of (possible)
Pyropoikilocytosis, genetic damage of
Pyrosis, genetic damage of
Pyruvate carboxylase deficiency, genetic damage of
Pyruvate decarboxylase deficiency, genetic damage of
Pyruvate dehydrogenase deficiency, genetic damage of
Pyruvate kinase deficiency, genetic damage of
Pyruvate kinase deficiency, liver type, genetic damage of
Pyruvate kinase deficiency, muscle type, genetic damage of

Q

Qazi Markouizos syndrome, genetic damage of
Quinquaud's decalvans folliculitis, genetic damage of

R

Rabson-Mendenhall syndrome, genetic damage of
Radial defect Robin sequence, genetic damage of
Radial deficiency tibial hypoplasia, genetic damage of
Radial hypoplasia triphalangeal thumbs hypospadias, genetic damage of
Radial ray agenesis, genetic damage of
Radial ray hypoplasia choanal atresia, genetic damage of
Radiation induced angiosarcoma of the breast, genetic damage of
Radiation leukemia, genetic damage of
Radiation related neoplasm /cancer, genetic damage of
Radiation syndromes, genetic damage of (possible)
Radiation-induced cancer, genetic damage of
Radicular dentin dysplasia, genetic damage of
Radiculomegaly of canine teeth congenital cataract, genetic damage of
Radio digito facial dysplasia, genetic damage of
Radio renal syndrome, genetic damage of
Radio-ulnar synostosis, genetic damage of
Radiophobia, genetic damage of (possible)
Radioulnar synostosis mental retardation hypotonia, genetic damage of
Radioulnar synostosis retinal pigment abnormalities, genetic damage of
Radius absent anogenital anomalies, genetic damage of
Raine syndrome, genetic damage of
Rambam Hasharon syndrome, genetic damage of
Rambaud Galian syndrome, genetic damage of
Ramer Ladda syndrome, genetic damage of
Ramon syndrome, genetic damage of
Ramos Arroyo Clark syndrome, genetic damage of
Ramsay Hunt paralysis syndrome, genetic damage of
Rapadilino syndrome, genetic damage of

Rapp-Hodgkin syndrome, genetic damage of
Rasmussen encephalitis, genetic damage of
Rasmussen Johnsen Thomsen syndrome, genetic damage of
Ray Peterson Scott syndrome, genetic damage of
Raynaud's disease/phenomenon, genetic damage of
Rayner Lampert Rennert syndrome, genetic damage of
Reactive arthritis, genetic damage of (possible)
Reactive attachment disorder of early childhood, genetic damage of (possible)
Reactive attachment disorder of infancy, genetic damage of (possible)
Reactive hypoglycemia, genetic damage of (possible)
Reardon Hall Slaney syndrome, genetic damage of
Reardon Wilson Cavanagh syndrome, genetic damage of
Rectal neoplasm, genetic damage of
Rectophobia, genetic damage of (possible)
Rectosigmoid neoplasm, genetic damage of
Recurrent laryngeal papillomas, genetic damage of
Recurrent peripheral facial palsy, genetic damage of
Reductional transverse limb defects, genetic damage of
Reflex sympathetic dystrophy syndrome, genetic damage of
Reflux esophagitis, genetic damage of
Refractory anemia, genetic damage of
Refsum disease, infantile form, genetic damage of
Refsum syndrome, genetic damage of
Reginato Shiapachasse syndrome, genetic damage of
Regional enteritis, genetic damage of
Reifenstein syndrome, genetic damage of
Reinhardt Pfeiffer syndrome, genetic damage of
Reiter's syndrome, genetic damage of
Renal adysplasia dominant type, genetic damage of
Renal agenesis, genetic damage of
Renal agenesis meningomyelocele mullerian defect, genetic dam-

age of
Renal agenesis, bilateral, genetic damage of
Renal artery stenosis, genetic damage of
Renal calculi, genetic damage of
Renal caliceal diverticuli deafness, genetic damage of
Renal cancer, genetic damage of
Renal carcinoma, familial, genetic damage of
Renal cell carcinoma, genetic damage of
Renal dysplasia diffuse autosomal recessive, genetic damage of
Renal dysplasia diffuse cystic, genetic damage of
Renal dysplasia hepatic fibrosis Dandy Walker cyst, genetic damage of
Renal dysplasia limb defects, genetic damage of
Renal dysplasia megalocystis sirenomelia, genetic damage of
Renal dysplasia mesomelia radiohumeral fusion, genetic damage of
Renal failure, genetic damage of
Renal genital middle ear anomalies, genetic damage of
Renal glycosuria, genetic damage of
Renal hepatic pancreatic dysplasia Dandy Walker cyst, genetic damage of
Renal hypertension, genetic damage of
Renal osteodystrophy, genetic damage of
Renal tubular acidosis, genetic damage of
Renal tubular acidosis progressive nerve deafness, genetic damage of
Renal tubular acidosis, distal, genetic damage of
Renal tubular acidosis, distal, autosomal dominant, genetic damage of
Renal tubular acidosis, distal, autosomal recessive, genetic damage of
Renal tubular acidosis, distal, type 3, genetic damage of
Renal tubular acidosis, distal, type 4, genetic damage of
Renal tubular transport disorders, genetic damage of

Rendu-Osler-Weber disease, genetic damage of
Renier Gabreels Jasper syndrome, genetic damage of
Renoanogenital syndrome, genetic damage of
Renoprival hypertension, genetic damage of
Resistance to LH (luteinizing hormone), genetic damage of
Resistance to thyroid stimulating hormone, genetic damage of
Respiratory chain deficiency malformations, genetic damage of
Respiratory distress syndrome, adult, genetic damage of
Respiratory distress syndrome, infant, genetic damage of
Restless legs syndrome, genetic damage of
Reticulosis, familial histiocytic, genetic damage of
Retina disorder, genetic damage of
Retinal degeneration, genetic damage of
Retinal dysplasia X linked, genetic damage of
Retinal telangiectasia hypogammaglobulinemia, genetic damage of
Retinis pigmentosa deafness hypogenitalism, genetic damage of
Retinitis pigmentosa, genetic damage of
Retinitis pigmentosa mental retardation deafness, genetic damage of
Retinitis pigmentosa-deafness, genetic damage of
Retinoblastoma, genetic damage of
Retinohepatoendocrinologic syndrome, genetic damage of
Retinopathy anemia CNS anomalies, genetic damage of
Retinopathy aplastic anemia neurological abnormalities, genetic damage of
Retinopathy pigmentary mental retardation, genetic damage of
Retinopathy, arteriosclerotic, genetic damage of
Retinopathy, diabetic, genetic damage of
Retinoschisis, genetic damage of
Retinoschisis, juvenile, genetic damage of
Retinoschisis, X linked, genetic damage of
Retraction syndrome, genetic damage of
Retrolental fibroplasia, genetic damage of

Retroperitoneal fibrosis, genetic damage of
Rett like syndrome, genetic damage of
Rett syndrome, genetic damage of
Revesz Debuse syndrome, genetic damage of
Reye syndrome, genetic damage of
Reynolds Neri Hermann syndrome, genetic damage of
Reynolds syndrome, genetic damage of
Rh disease, genetic damage of
Rhabdoid tumor, genetic damage of
Rhabdomyomatous dysplasia cardiopathy genital anomalies, genetic damage of
Rhabdomyosarcoma, genetic damage of
Rhabdomyosarcoma 1, genetic damage of
Rhabdomyosarcoma 2, genetic damage of
Rhabdomyosarcoma, alveolar, genetic damage of
Rhabdomyosarcoma, embryonal, genetic damage of
Rheumatic Fever, genetic damage of (possible)
Rheumatism, genetic damage of
Rheumatoid arthritis, genetic damage of
Rheumatoid vasculitis, genetic damage of
Rhizomelic dysplasia type Patterson Lowry, genetic damage of
Rhizomelic pseudopolyarthritis, genetic damage of
Rhizomelic syndrome, genetic damage of
Rhumatoid purpura, genetic damage of
Rhypophobia, genetic damage of (possible)
Rhytiphobia, genetic damage of (possible)
Richieri Costa Colletto Otto syndrome, genetic damage of
Richieri Costa Da Silva syndrome, genetic damage of
Richieri Costa Gorlin syndrome, genetic damage of
Richieri Costa Guion Almeida acrofacial dysostosis, genetic damage of
Richieri Costa Guion Almeida Cohen syndrome, genetic damage of

Richieri Costa Guion Almeida dwarfism, genetic damage of
Richieri Costa Guion Almeida Rodini syndrome, genetic damage of
Richieri Costa Guion Almeida syndrome, genetic damage of
Richieri Costa Montagnoli syndrome, genetic damage of
Richieri Costa Orquizas syndrome, genetic damage of
Richieri Costa Silveira Pereira syndrome, genetic damage of
Richter syndrome, genetic damage of
Rieger syndrome, genetic damage of
Right atrium familial dilatation, genetic damage of
Right ventricle hypoplasia, genetic damage of
Rigid mask like face deafness polydactyly, genetic damage of
Rigid spine syndrome, genetic damage of
Riley-Day syndrome, genetic damage of
Ring chromosome 17, genetic damage of
Ringed hair disease, genetic damage of
Rippberger Aase syndrome, genetic damage of
Rivera Perez Salas syndrome, genetic damage of
Roberts syndrome, genetic damage of
Robin sequence oligodactyly, genetic damage of
Robinow Sorauf syndrome, genetic damage of
Robinow syndrome, genetic damage of
Robinow syndrome recessive form, genetic damage of
Robinson Miller Bensimon syndrome, genetic damage of
Roch-Leri mesosomatous lipomatosis, genetic damage of
Rod myopathy, genetic damage of
Rodini Richieri Costa syndrome, genetic damage of
Rokitansky Kuster Hauser syndrome, genetic damage of
Rokitansky sequence, genetic damage of
Romano-Ward syndrome, genetic damage of
Romberg hemi-facial atrophy, genetic damage of
Rombo syndrome, genetic damage of
Rommen Mueller Sybert syndrome, genetic damage of

Rosai-Dorfman disease, genetic damage of
Rosenberg Chutorian syndrome, genetic damage of
Rosenberg Lohr syndrome, genetic damage of
Rothmund-Thomson syndrome, genetic damage of
Rotor syndrome, genetic damage of
Roussy Levy hereditary areflexic dystasia, genetic damage of
Roussy-Levy syndrome, genetic damage of
Roy Maroteaux Kremp syndrome, genetic damage of
Rozin Hertz Goodman syndrome, genetic damage of
Rubinstein Taybi like syndrome, genetic damage of
Rubinstein-Taybi syndrome, genetic damage of
Rudd Klimek syndrome, genetic damage of
Rudiger syndrome, genetic damage of
Rumination disorder, genetic damage of (possible)
Rupophobia, genetic damage of (possible)
Rutledge Friedman Harrod syndrome, genetic damage of
Ruvalcaba Churesigaew Myhre syndrome, genetic damage of
Ruvalcaba syndrome, genetic damage of
Ruvalcaba-Myhre syndrome, genetic damage of
Ruvalcaba-Myhre-Smith syndrome (BRR), genetic damage of
Ruzicka Goerz Anton syndrome, genetic damage of

S

Saal Bulas syndrome, genetic damage of
Saal Greenstein syndrome, genetic damage of
Sabinas brittle hair syndrome, genetic damage of
Saccharopinuria, genetic damage of
Sackey Sakati Aur syndrome, genetic damage of
Sacral agenesis, genetic damage of
Sacral defect anterior sacral meningocele, genetic damage of
Sacral hemangiomas multiple congenital abnormaliti, genetic damage of
Sacral meningocele conotruncal heart defects, genetic damage of
Sacrococcygeal dysgenesis association, genetic damage of
Saethre-Chotzen syndrome, genetic damage of
Saito Kuba Tsuruta syndrome, genetic damage of
Sakati syndrome, genetic damage of
Salcedo syndrome, genetic damage of
Saldino Mainzer syndrome, genetic damage of
Salivary disorder, genetic damage of
Salivary gland disorders, genetic damage of
Salla disease, genetic damage of
Sallis Beighton syndrome, genetic damage of
Salti Salem syndrome, genetic damage of
Sammartino Decreccio syndrome, genetic damage of
Samson Gardner syndrome, genetic damage of
Samson Viljoen syndrome, genetic damage of
Sanderson Fraser syndrome, genetic damage of
Sandhaus Ben Ami syndrome, genetic damage of
Sandhoff disease, genetic damage of
Sanfilippo syndrome, genetic damage of
Sanfilippo syndrome type A, genetic damage of
Sanfilippo syndrome type B, genetic damage of
Sanfilippo syndrome type C, genetic damage of
Sanfilippo syndrome type D, genetic damage of

Santavuori disease, genetic damage of
Santavuori-Haltia disease, genetic damage of
Santos Mateus Leal syndrome, genetic damage of
SAPHO syndrome, genetic damage of
Sarcoidosis, genetic damage of
Sarcoidosis, pulmonary, genetic damage of
Sarcosinemia, genetic damage of
Satoyoshi syndrome, genetic damage of
Saul Wilkes Stevenson syndrome, genetic damage of
Say Barber Hobbs syndrome, genetic damage of
Say Barber Miller syndrome, genetic damage of
Say Carpenter syndrome, genetic damage of
Say Field Coldwell syndrome, genetic damage of
Say Meyer syndrome, genetic damage of
Scabiophobia, genetic damage of (possible)
SCAD deficiency, genetic damage of
Scalp defects postaxial polydactyly, genetic damage of
Scalp ear nipple syndrome, genetic damage of
Scapuloiliac dysostosis, genetic damage of
Scapuloperoneal myopathy, genetic damage of
Scarf syndrome, genetic damage of
Scatophobia, genetic damage of (possible)
Schaap Taylor Baraitser syndrome, genetic damage of
Schaefer Stein Oshman syndrome, genetic damage of
Schamberg disease, genetic damage of
Schamberg disease pigmentation disorder, genetic damage of
Scheie syndrome, genetic damage of
Schereshevskij Turner, genetic damage of
Scheurermann's disease, genetic damage of
Schimke syndrome, genetic damage of
Schindler disease, genetic damage of
Schinzel acrocallosal syndrome, genetic damage of
Schinzel Giedion syndrome, genetic damage of

Schinzel syndrome, genetic damage of
Schinzel-Giedion midface retraction syndrome, genetic damage of
Schisis association, genetic damage of
Schizencephaly, genetic damage of
Schizophrenia mental retardation deafness retinitis, genetic damage of
Schizophrenia, genetic types, genetic damage of
Schlegelberger Grote syndrome, genetic damage of
Schmidt syndrome, genetic damage of
Schmitt Gillenwater Kelly syndrome, genetic damage of
Schneckenbecken dysplasia, genetic damage of
Schofer Beetz Bohl syndrome, genetic damage of
Scholte Begeer Van Essen syndrome, genetic damage of
Schonlein-Henoch purpura, genetic damage of
Schraderman's disease, genetic damage of
Schrander Stumpel Theunissen Hulsmans syndrome, genetic damage of
Schroer Hammer Mauldin syndrome, genetic damage of
Schwannoma, malignant, genetic damage of
Schwannomatosis, genetic damage of
Schwartz Newark syndrome, genetic damage of
Schwartz-Jampel syndrome, genetic damage of
Schweitzer Kemink Malcolm syndrome, genetic damage of
Scimitar syndrome, genetic damage of
Sciophobia, genetic damage of (possible)
Scleroatonic myopathy, genetic damage of
Sclerocornea syndactyly ambiguous genitalia, genetic damage of
Scleroderma, genetic damage of
Scleromyxedema, genetic damage of
Sclerosing bone dysplasia mental retardation, genetic damage of
Sclerosing cholangitis, genetic damage of
Sclerosing Mesenteritis, genetic damage of
Sclerosteosis, genetic damage of

Scoditti Geminiani Colonna syndrome, genetic damage of
Scoleciphobia, genetic damage of (possible)
Scoliosis as part of NF, genetic damage of
Scoliosis with unilateral unsegmented bar, genetic damage of
Scopophobia, genetic damage of (possible)
SCOT deficiency, genetic damage of
Scotomaphobia, genetic damage of (possible)
Scott Aarskog syndrome, genetic damage of
Scott Bryant Graham syndrome, genetic damage of
Scott craniodigital syndrome with mental retardation, genetic damage of
Scott syndrome, genetic damage of
Sea-blue histiocytosis, genetic damage of
Seaver Cassidy syndrome, genetic damage of
Sebocystomatosis, genetic damage of
Seborrheic keratosis, genetic damage of
Seckel like syndrome Majoor Krakauer type, genetic damage of
Seckel like syndrome type Buebel, genetic damage of
Seckel syndrome, genetic damage of
Secondary pulmonary hypertension, genetic damage of
Seemanova Lesny syndrome, genetic damage of
Seemanova syndrome type 2, genetic damage of
Segawa syndrome, genetic damage of
Seghers syndrome, genetic damage of
Segmental neurofibromatosis, genetic damage of
Segmental vertebral anomalies, genetic damage of
Seitelberger disease, genetic damage of
Seizures benign familial neonatal recessive form, genetic damage of
Seizures mental retardation hair dysplasia, genetic damage of
Selachophobia, genetic damage of (possible)
Selenophobia, genetic damage of (possible)
Selig Benacerraf Greene syndrome, genetic damage of
Seminoma, genetic damage of

Semmerkrot Haraldsson Weenaes syndrome, genetic damage of
Sengers Hamel Otten syndrome, genetic damage of
Senior syndrome, genetic damage of
Sensorineural hearing loss, genetic damage of (possible)
Sensory neuropathy, genetic damage of
Sensory neuropathy type 1, genetic damage of
Sensory radicular neuropathy recessive form, genetic damage of
Senter syndrome, genetic damage of
Seow Najjar syndrome, genetic damage of
Seplophobia, genetic damage of (possible)
Septo-optic dysplasia, genetic damage of
Septooptic dysplasia, genetic damage of
Septooptic dysplasia digital anomalies, genetic damage of
Sequeiros Sack syndrome, genetic damage of
Seres Santamaria Arimany Muniz syndrome, genetic damage of
Setleis syndrome, genetic damage of
Severe combined immunodeficiency, genetic damage of
Severe combined immunodeficiency, alymphocytotic type, genetic damage of
Severe combined immunodeficiency, HLA class 2-negative, genetic damage of
Severe infantile axonal neuropathy, genetic damage of
Sexual precocity, familial, gonadotropin-independent, genetic damage of
Sezary syndrome, genetic damage of
Sezary's lymphoma, genetic damage of
Shapiro syndrome, genetic damage of
Sharma Kapoor Ramji syndrome, genetic damage of
Sharp syndrome, genetic damage of
Sheehan syndrome, genetic damage of
Shith Filkins syndrome, genetic damage of
Shokeir syndrome, genetic damage of
Short broad great toe macrocranium, genetic damage of

Short chain Acyl CoA dehydrogenase deficiency, genetic damage of

Short limb dwarf lethal Colavita Kozlowski type, genetic damage of

Short limb dwarf lethal Mcalister Crane type, genetic damage of

Short limb dwarf mental retardation myopia, genetic damage of

Short limb dwarf oedema iris coloboma, genetic damage of

Short limb dwarfism Al Gazali type, genetic damage of

Short limbs abnormal face congenital heart disease, genetic damage of

Short limbs subluxed knees cleft palate, genetic damage of

Short rib syndrome Beemer type, genetic damage of

Short rib-polydactyly syndrome, genetic damage of

Short rib-polydactyly syndrome, Beermer type, genetic damage of

Short rib-polydactyly syndrome, Majewski type, genetic damage of

Short rib-polydactyly syndrome, Saldino-Noonan type, genetic damage of

Short rib-polydactyly syndrome, Verma-Naumoff type, genetic damage of

Short ribs craniosynostosis polysyndactyly, genetic damage of

Short stature abnormal skin pigmentation mental retardation, genetic damage of

Short stature Brussels type, genetic damage of

Short stature contractures hypotonia, genetic damage of

Short stature cranial hyperostosis hepatomegaly, genetic damage of

Short stature deafness neutrophil dysfunction, genetic damage of

Short stature dysmorphic face pelvic scapula dysplasia, genetic damage of

Short stature heart defect craniofacial anomalies, genetic damage of

Short stature hyperkaliemia acidosis, genetic damage of

Short stature locking fingers, genetic damage of

Short stature mental retardation eye anomalies, genetic damage of

Short stature mental retardation eye defects, genetic damage of
Short stature microcephaly heart defect, genetic damage of
Short stature microcephaly seizures deafness, genetic damage of
Short stature monodactylous ectrodactyly cleft palate, genetic damage of
Short stature prognathism short femoral necks, genetic damage of
Short stature Robin sequence cleft mandible hand anomalies club-foot, genetic damage of
Short stature talipes natal teeth, genetic damage of
Short stature valvular heart disease, genetic damage of
Short stature webbed neck heart disease, genetic damage of
Short stature wormian bones dextrocardia, genetic damage of
Short syndrome, genetic damage of
Short tarsus absence of lower eyelashes, genetic damage of
Shoulder and thorax deformity congenital heart disease, genetic damage of
Shoulder girdle defect mental retardation familial, genetic damage of
Shprintzen Golberg craniosynostosis, genetic damage of
Shprintzen syndrome, genetic damage of
Shulman syndrome, genetic damage of
Shwachman syndrome, genetic damage of
Shwachman-Diamond syndrome, genetic damage of
Shy-Drager syndrome, genetic damage of
Sialadenitis, genetic damage of
Sialidosis, genetic damage of
Sialidosis type 1 and 3, genetic damage of
Sialuria french type, genetic damage of
Sickle cell anemia, genetic damage of
Sickle cell crisis, genetic damage of
Sickle cell trait, genetic damage of
Sideroblastic anemia, genetic damage of
Siderodromophobia, genetic damage of (possible)

Sidransky Feinstein Goodman syndrome, genetic damage of
Siegler Brewer Carey syndrome, genetic damage of
Silengo Lerone Pelizzo syndrome, genetic damage of
Sillence syndrome, genetic damage of
Silver-Russell dwarfism, genetic damage of
Silvery hair syndrome, genetic damage of
Simosa Penchaszadeh Bustos syndrome, genetic damage of
Simpson-Golabi-Behmel syndrome, genetic damage of
Singh Chhaparwal Dhanda syndrome, genetic damage of
Single upper central incisor, genetic damage of
Single ventricular heart, genetic damage of
Singleton Merten syndrome, genetic damage of
Sinistrophobia, genetic damage of (possible)
Sino-auricular heart block, genetic damage of
Sinus cancer, genetic damage of
Sinus histiocytosis, genetic damage of
Sinus node disease and myopia, genetic damage of
Sipple syndrome, genetic damage of
Sirenomelia, genetic damage of
Sirenomelia sequence, genetic damage of
Sitophobia, genetic damage of (possible)
Sitosterolemia, genetic damage of
Situs inversus viscerum-cardiopathy, genetic damage of
Situs inversus, X linked, genetic damage of
Sjogren Larsson like syndrome, genetic damage of
Sjogren Larsson syndrome, genetic damage of
Sjogren's syndrome, genetic damage of
Skeletal dysplasia brachydactyly, genetic damage of
Skeletal dysplasia epilepsy short stature, genetic damage of
Skeletal dysplasia orofacial anomalies, genetic damage of
Skeletal dysplasia San Diego type, genetic damage of
Skeletal dysplasias, genetic damage of
Skeleto cardiac syndrome with thrombocytopenia, genetic damage

of
Sketetal dysplasia coarse facies mental retardation, genetic damage
of
Slavotinek Hurst syndrome, genetic damage of
Sleep apnea, genetic damage of
Sly syndrome, genetic damage of
Small cell lung cancer, genetic damage of
Small non-cleaved cell lymphoma, genetic damage of
Small patella syndrome, genetic damage of
Smet Fabry Fryns syndrome, genetic damage of
Smith Fineman Myers syndrome, genetic damage of
Smith Martin Dodd syndrome, genetic damage of
Smith-Magenis syndrome, genetic damage of
Sneddon syndrome, genetic damage of
Sociophobia, genetic damage of (possible)
Soft tissue sarcomas, genetic damage of
Sohval Soffer syndrome, genetic damage of
Somatostatinoma, genetic damage of
Sommer Hines syndrome, genetic damage of
Sommer Rathbun Battles syndrome, genetic damage of
Sommer Young Wee Frye syndrome, genetic damage of
Somniphobia, genetic damage of (possible)
Sondheimer syndrome, genetic damage of
Sonoda syndrome, genetic damage of
Sophophobia, genetic damage of (possible)
Sosby syndrome, genetic damage of
Sotos syndrome, genetic damage of
Sparse hair ptosis mental retardation, genetic damage of
Spasmodic dysphonia, genetic damage of
Spasmodic torticollis, genetic damage of
Spastic angina with healthy coronary artery, genetic damage of
Spastic ataxia Charlevoix-Saguenay type, genetic damage of
Spastic diplegia infantile type, genetic damage of

Spastic dysphonia, genetic damage of
Spastic paraparesis, genetic damage of
Spastic paraparesis deafness, genetic damage of
Spastic paraparesis, infantile, genetic damage of
Spastic paraplegia epilepsy mental retardation, genetic damage of
Spastic paraplegia facial cutaneous lesions, genetic damage of
Spastic paraplegia familial autosomal recessive form, genetic damage of
Spastic paraplegia glaucoma precocious puberty, genetic damage of
Spastic paraplegia mental retardation corpus callosum, genetic damage of
Spastic paraplegia nephritis deafness, genetic damage of
Spastic paraplegia neuropathy poikiloderma, genetic damage of
Spastic paraplegia type 1, X linked, genetic damage of
Spastic paraplegia type 2, X linked, genetic damage of
Spastic paraplegia type 3, dominant, genetic damage of
Spastic paraplegia type 4, dominant, genetic damage of
Spastic paraplegia type 5A, recessive, genetic damage of
Spastic paraplegia type 5B, recessive, genetic damage of
Spastic paraplegia type 6, dominant, genetic damage of
Spastic paraplegia, familial, genetic damage of
Spastic paresis glaucoma mental retardation, genetic damage of
Spastic quadriplegia retinitis pigmentosa mental retardation, genetic damage of
Spasticity mental retardation, genetic damage of
Spasticity multiple exostoses, genetic damage of
Spatic paraparesis vitiligo premature graying, genetic damage of
Spellacy Gibbs Watts syndrome, genetic damage of
Spherocytosis, genetic damage of
Spherophakia brachymorphia syndrome, genetic damage of
Sphingolipidosis, genetic damage of
Sphingomyelinase deficiency, genetic damage of
Spielmeyer-Vogt disease, genetic damage of

Spina bifida, genetic damage of
Spina bifida hypospadias, genetic damage of
Spinal and bulbar muscular atrophy, genetic damage of
Spinal atrophy ophthalmoplegia pyramidal syndrome, genetic damage of
Spinal bulbar motor neuropathy, genetic damage of
Spinal bulbar muscular atrophy, genetic damage of
Spinal cord disorder, genetic damage of
Spinal cord neoplasm, genetic damage of
Spinal dysostosis type Anhalt, genetic damage of
Spinal muscular atrophy, genetic damage of
Spinal muscular atrophy type 1, genetic damage of
Spinal muscular atrophy type 2, genetic damage of
Spinal muscular atrophy type 3, genetic damage of
Spinal muscular atrophy type I with congenital bone fractures, genetic damage of
Spinal stenosis, genetic damage of
Spine rigid cardiomyopathy, genetic damage of
Spinocerebellar ataxia 1, genetic damage of
Spinocerebellar ataxia 2, genetic damage of
Spinocerebellar ataxia 4, genetic damage of
Spinocerebellar ataxia 5, genetic damage of
Spinocerebellar ataxia 6, genetic damage of
Spinocerebellar ataxia 7, genetic damage of
Spinocerebellar ataxia 8, genetic damage of
Spinocerebellar ataxia amyotrophy deafness, genetic damage of
Spinocerebellar ataxia dysmorphism, genetic damage of
Spinocerebellar atrophy type 3, genetic damage of
Spinocerebellar degeneration corneal dystrophy, genetic damage of
Spinocerebellar degenerescence book type, genetic damage of
Spleen neoplasm, genetic damage of
Splenic agenesis syndrome, genetic damage of
Splenogonadal fusion limb defects micrognatia, genetic damage of

Splenomegaly, genetic damage of

Split hand deformity mandibulofacial dysostosis, genetic damage of

Split hand split foot malformation autosomal reces, genetic damage of

Split hand split foot mandibular hypoplasia, genetic damage of

Split hand split foot nystagmus, genetic damage of

Split hand split foot X linked, genetic damage of

Split hand urinary anomalies spina bifida, genetic damage of

Split-hand deformity, genetic damage of

Sponastrime dysplasia, genetic damage of

Spondylarthropathy, genetic damage of

Spondylitis, genetic damage of

Spondylo camptodactyly syndrome, genetic damage of

Spondylo costal dysostosis dandy walker, genetic damage of

Spondylocarpotarsal synostosis, genetic damage of

Spondylocostal dysplasia dominant, genetic damage of

Spondylodysplasia brachyolmia, genetic damage of

Spondyloenchondrodysplasia, genetic damage of

Spondyloepimetaphyseal dysplasia, genetic damage of

Spondyloepimetaphyseal dysplasia congenita, Iraqi type, genetic damage of

Spondyloepimetaphyseal dysplasia congenita, Strudwick type, genetic damage of

Spondyloepimetaphyseal dysplasia joint laxity, genetic damage of

Spondyloepiphyseal dysplasia, genetic damage of

Spondyloepiphyseal dysplasia nephrotic syndrome, genetic damage of

Spondyloepiphyseal dysplasia tarda, genetic damage of

Spondyloepiphyseal dysplasia tarda progressive art, genetic damage of

Spondyloepiphyseal dysplasia, congenital type, genetic damage of

Spondylohypoplasia arthrogryposis popliteal pterygium, genetic

damage of
Spondylometaphyseal dysplasia, genetic damage of
Spondylometaphyseal dysplasia Kozlowski type, genetic damage of
Spondylometaphyseal dysplasia, 'corner fracture' type, genetic damage of
Spondylometaphyseal dysplasia, Algerian type, genetic damage of
Spondylometaphyseal dysplasia, Schmidt type, genetic damage of
Spondylometaphyseal dysplasia, Sedaghatian type, genetic damage of
Spondyloperipheral dysplasia short ulna, genetic damage of
Spongy degeneration of central nervous system, genetic damage of
Spontaneous periodic hypothermia, genetic damage of (possible)
Spranger Schinzel Yers syndrome, genetic damage of
Sprengel deformity, genetic damage of
Squamous cell carcinoma, genetic damage of
SSADH (succinic semialdehyde dehydrogenase deficiency), genetic damage of
Stalker Chitayat syndrome, genetic damage of
Stampe Sorensen syndrome, genetic damage of
Stargardt's disease, genetic damage of
Steatocystoma multiplex, genetic damage of
Steatocystoma multiplex natal teeth, genetic damage of
Steele Richardson Olszewski syndrome atypical, genetic damage of
Stein-Leventhal syndrome, genetic damage of
Steinbrocker syndrome, genetic damage of
Steinert disease, genetic damage of
Steinert myotonic dystrophy, genetic damage of
Steinfeld syndrome, genetic damage of
Stenophobia, genetic damage of (possible)
Stern Lubinsky Durrie syndrome, genetic damage of
Sternal cleft, genetic damage of
Sternal cyst vascular anomalies, genetic damage of
Sternal malformation vascular dysplasia associatio, genetic damage

of

Steroid dehydrogenase deficiency dental anomalies, genetic damage of

Steroid sulfatase deficiency, genetic damage of

Stevens-Johnson syndrome, genetic damage of

Stickler syndrome, genetic damage of

Stickler syndrome, type 1, genetic damage of

Stickler syndrome, type 2, genetic damage of

Stickler syndrome, type 3, genetic damage of

Stiff man syndrome, genetic damage of

Stiff skin syndrome, genetic damage of

Still's disease, genetic damage of

Stimmler syndrome, genetic damage of

Stoelinga de Koomen Davis syndrome, genetic damage of

Stoll Alembik Dott syndrome, genetic damage of

Stoll Alembik Finck syndrome, genetic damage of

Stoll Geraudel Chauvin syndrome, genetic damage of

Stoll Kieny Dott syndrome, genetic damage of

Stoll Levy Francfort syndrome, genetic damage of

Stomach cancer, genetic damage of

Stomach cancer, familial, genetic damage of

Storage pool platelet disease, genetic damage of

Stormorken Sjaastad Langslet syndrome, genetic damage of

Strabismus, genetic damage of

Stratton Garcia Young syndrome, genetic damage of

Stratton Parker syndrome, genetic damage of

Streeter's (Amniotic bands), genetic damage of

Striatal degeneration familial, genetic damage of

Striatonigral degeneration infantile, genetic damage of

Strudwick syndrome, genetic damage of

Strumpell-Lorrain disease, genetic damage of

Stuart factor deficiency, congenital, genetic damage of

Stucco keratosis, genetic damage of

Sturge-Weber syndrome, genetic damage of
Stuve Wiedemann dysplasia, genetic damage of
Subacute cerebellar degeneration, genetic damage of
Subacute sclerosing leucoencephalitis, genetic damage of
Subacute sclerosing panencephalitis, genetic damage of
Subaortic stenosis short stature syndrome, genetic damage of
Subcortical laminar heterotopia, genetic damage of
Subependymal nodular heterotopia, genetic damage of
Subpulmonary stenosis, genetic damage of
Subvalvular aortic stenosis, genetic damage of
Succinate coenzyme Q reductase deficiency of, genetic damage of
Succinic acidemia, genetic damage of
Succinic acidemia lactic acidosis congenital, genetic damage of
Succinic semialdehyde dehydrogenase deficiency, genetic damage of
Succinyl-CoA acetoacetate transferase deficiency, genetic damage of
Sucrase-isomaltase deficiency, genetic damage of
Sudden infant death syndrome, genetic damage of
Sugarman syndrome, genetic damage of
Sulfatidosis juvenile, Austin type, genetic damage of
Sulfite and xanthine oxydase deficiency, genetic damage of
Sulfite oxidase deficiency, genetic damage of
Summitt syndrome, genetic damage of
Super mesenteric artery syndrome, genetic damage of
Supranuclear palsy, progressive, genetic damage of
Suriphobia, genetic damage of (possible)
Susac syndrome, genetic damage of
Sutherland Haan syndrome, genetic damage of
Sutton's disease II, genetic damage of
Sweet syndrome, genetic damage of
Swyer syndrome, genetic damage of
Sybert Smith syndrome, genetic damage of

Symmetrical thalamic calcifications, genetic damage of
Symphalangism brachydactyly, genetic damage of
Symphalangism brachydactyly craniosynostosis, genetic damage of
Symphalangism Cushing type, genetic damage of
Symphalangism distal, genetic damage of
Symphalangism familial proximal, genetic damage of
Symphalangism short stature accessory testis, genetic damage of
Symphalangism with multiple anomalies of hands and feet, genetic damage of
Syncamptodactyly scoliosis, genetic damage of
Syncopal paroxysmal tachycardia, genetic damage of
Syncopal tachyarythmia, genetic damage of
Syndactyly, genetic damage of
Syndactyly between 4 and 5, genetic damage of
Syndactyly cataract mental retardation, genetic damage of
Syndactyly Cenani Lenz type, genetic damage of
Syndactyly ectodermal dysplasia cleft lip palate hand foot, genetic damage of
Syndactyly type 1 microcephaly mental retardation, genetic damage of
Syndactyly type 2, genetic damage of
Syndactyly type 3, genetic damage of
Syndactyly type 5, genetic damage of
Syndactyly-polydactyly-ear lobe syndrome, genetic damage of
Syndrome X, genetic damage of
Syngnathia cleft palate, genetic damage of
Syngnathia multiple anomalies, genetic damage of
Synostosis of talus and calcaneus short stature, genetic damage of
Synovial cancer, genetic damage of
Synovial osteochondromatosis, genetic damage of
Synovial sarcoma, genetic damage of
Synovitis, genetic damage of
Synovitis acne pustulosis hyperostosis osteitis, genetic damage of

Synovitis granulomatous uveitis cranial neuropathi, genetic damage of
Synpolydactyly, genetic damage of
Synspondylism, genetic damage of
Syringobulbia, genetic damage of
Syringocystadenoma papilliferum, genetic damage of
Syringomas natal teeth oligodontia, genetic damage of
Syringomelia hyperkeratosis, genetic damage of
Syringomyelia, genetic damage of
Systemic arterio-veinous fistula, genetic damage of
Systemic carnitine deficiency, genetic damage of
Systemic lupus erythematosus, genetic damage of
Systemic mastocytosis, genetic damage of
Systemic sclerosis, genetic damage of

T

T cell immunodeficiency primary, genetic damage of
T-cell lymphoma, genetic damage of
T-Lymphocytopenia, genetic damage of
Tabatznik syndrome, genetic damage of
Tachycardia, genetic damage of
Taeniophobia, genetic damage of (possible)
Takayasu arteritis, genetic damage of
Talipes equinovarus, genetic damage of
Tamari Goodman syndrome, genetic damage of
Tang Hsi Ryu syndrome, genetic damage of
Tangier disease, genetic damage of
Tapinophobia, genetic damage of (possible)
Tar syndrome, genetic damage of
Tardive dyskinesia, genetic damage of
Tarsal tunnel syndrome, genetic damage of
Taste disorder, genetic damage of
Tau syndrome, genetic damage of
Taurodontia absent teeth sparse hair, genetic damage of
Taurodontism, genetic damage of
Tay syndrome ichthyosis, genetic damage of
Tay-Sachs disease, genetic damage of
Taybi Linder syndrome, genetic damage of
Taybi syndrome, genetic damage of
Teebi Kaurah syndrome, genetic damage of
Teebi Naguib Alawadi syndrome, genetic damage of
Teebi Shaltout syndrome, genetic damage of
Teebi syndrome, genetic damage of
Teeth noneruption of with maxillary hypoplasia and genu valgum, genetic damage of
Tel Hashomer camptodactyly syndrome, genetic damage of
Telangiectasia, genetic damage of
Telangiectasia ataxia variant V1, genetic damage of

Telangiectasia, hereditary hemorrhagic, genetic damage of
Telecanthus hypertelorism pes cavus, genetic damage of
Telecanthus with associated abnormalities, genetic damage of
Telencephalic leukoencephalopathy, genetic damage of
Telfer Sugar Jaeger syndrome, genetic damage of
Temporal epilepsy, familial, genetic damage of
Temporomandibular ankylosis, genetic damage of
Temporomandibular joint dysfunction, genetic damage of
Temtamy Shalash syndrome, genetic damage of
TEN, genetic damage of (possible)
Ter Haar Hamel Hendricks syndrome, genetic damage of
Ter Haar syndrome, genetic damage of
Teratocarcinosarcoma, genetic damage of
Teratoma, genetic damage of
Teratophobia, genetic damage of (possible)
Testes neoplasm, genetic damage of
Testicular feminization syndrome, genetic damage of
Testicular regression syndrome, genetic damage of
Testotoxicosis, genetic damage of
Tetanophobia, genetic damage of (possible)
Tethered spinal cord disease, genetic damage of
Tetraamelia ectodermal dysplasia, genetic damage of
Tetraamelia multiple malformations, genetic damage of
Tetraamelia pulmonary hypoplasia, genetic damage of
Tetraamelia-syrinx, genetic damage of
Tetrahydrobiopterin deficiencies, genetic damage of
Tetraploidy, genetic damage of
Tetrasomy 12p, genetic damage of
Tetrasomy 15q, genetic damage of
Tetrasomy 18p, genetic damage of
Tetrasomy 21q, genetic damage of
Tetrasomy 9p, genetic damage of
Tetrasomy X, genetic damage of

Thaasophobia, genetic damage of (possible)
Thakker Donnai syndrome, genetic damage of
Thalamic degeneration symmetrical infantile, genetic damage of
Thalamic degenerescence infantile, genetic damage of
Thalamic syndrome, genetic damage of
Thalassemia, genetic damage of
Thalassemia major, genetic damage of
Thalassemia minor, genetic damage of
Thalassophobia, genetic damage of (possible)
Thanatophobia, genetic damage of (possible)
Thanatophoric dwarfism, genetic damage of
Thanatophoric dysplasia cloverleaf skull, genetic damage of
Thanatophoric dysplasia Glasgow variant, genetic damage of
Thanos Stewart Zonana syndrome, genetic damage of
Theodor Hertz Goodman syndrome, genetic damage of
Thiele syndrome, genetic damage of
Thiemann epiphyseal disease, genetic damage of
Thies Reis syndrome, genetic damage of
Thin ribs tubular bones dysmorphism, genetic damage of
Thiolase deficiency, genetic damage of
Thiopurine S methyltranferase deficiency, genetic damage of
Thomas Jewett Raines syndrome, genetic damage of
Thomas syndrome, genetic damage of
Thombocytopenia X linked, genetic damage of
Thompson Baraitser syndrome, genetic damage of
Thong Douglas Ferrante syndrome, genetic damage of
Thoracic celosomia, genetic damage of
Thoracic dysplasia hydrocephalus syndrome, genetic damage of
Thoracic outlet syndrome, genetic damage of
Thoraco abdominal enteric duplication, genetic damage of
Thoraco limb dysplasia Rivera type, genetic damage of
Thoracolaryngopelvic dysplasia, genetic damage of
Thoracopelvic dysostosis, genetic damage of

Thyroid renal digital anomalies, genetic damage of
Tibia absent polydactyly, genetic damage of
Tibia absent polydactyly arachnoid cyst, genetic damage of
Tibiae bowed radial anomalies osteopennia fracture, genetic damage of
Tibial aplasia ectrodactyly, genetic damage of
Tibial aplasia ectrodactyly hydrocephalus, genetic damage of
Tibial hemimelia cleft lip palate, genetic damage of
Tibial muscular dystrophy tardive, genetic damage of
Tietze syndrome, genetic damage of
Tinnitus, genetic damage of
TNF-1 receptor associated progressive syndrome, genetic damage of
Todd's paralysis, genetic damage of
Tollner Horst Manzke syndrome, genetic damage of
Tolosa-Hunt syndrome, genetic damage of
Tomaculous neuropathy, genetic damage of
Tome Brune Fardeau syndrome, genetic damage of
Tongue neoplasm, genetic damage of
Toni Debre Fanconi maladie, genetic damage of
Toni-Fanconi syndrome, genetic damage of
Topophobia, genetic damage of (possible)
Toriello Carey syndrome, genetic damage of
Toriello Lacassie Droste syndrome, genetic damage of
Toriello syndrome, genetic damage of
Toriello-Higgins-Miller syndrome, genetic damage of
Torres Ayber syndrome, genetic damage of
Torsion dystonia, genetic damage of
Torticollis keloids cryptorchidism renal dysplasia, genetic damage of
Tosti Misciali Barbareschi syndrome, genetic damage of
Total hypotrichosis, Mari type, genetic damage of
Tourette syndrome, genetic damage of

Townes-Brocks syndrome, genetic damage of
Toxic encephalopathy, genetic damage of
Toxopachyoteose diaphysaire tibio peroniere, genetic damage of
Tracheal agenesis, genetic damage of
Tracheobronchomalacia, genetic damage of
Tracheobronchomegaly, genetic damage of
Tracheobronchopathia osteoplastica, genetic damage of
Tracheoesophageal fistula, genetic damage of
Tracheoesophageal fistula symphalangism, genetic damage of
Tracheophageal fistula hypospadias, genetic damage of
Tranebjaerg Svejgaard syndrome, genetic damage of
Transcobalamin II deficiency, genetic damage of
Transient erythroblastopenia of childhood, genetic damage of
Transient global amnesia, genetic damage of
Transient neonatal arthrogryposis, genetic damage of
Transitional cell carcinoma, genetic damage of
Transposition of great vessels, genetic damage of
Transverse limb deficiency hemangioma, genetic damage of
Transverse myelitits, genetic damage of
TRAPS (TNF-receptor-associated periodic syndrome), genetic damage of
Traumatophobia, genetic damage of (possible)
Treacher Collins-Franceschetti syndrome, genetic damage of
Treft Sanborn Carey syndrome, genetic damage of
Tremophobia, genetic damage of (possible)
Tremor hereditary essential, genetic damage of
Tremor nystagmus duodenal ulcer, genetic damage of
Trevor disease, genetic damage of
Triatrial heart, genetic damage of
Tricho dento osseous syndrome type 1, genetic damage of
Tricho odonto onycho dermal syndrome, genetic damage of
Tricho odonto onychodysplasia syndactyly dominant type, genetic damage of

Tricho onychic dysplasia, genetic damage of
Tricho onycho hypohidrotic dysplasia, genetic damage of
Tricho retino dento digital syndrome, genetic damage of
Tricho-dento-osseous syndrome, genetic damage of
Tricho-hepato-enteric syndrome, genetic damage of
Trichodental syndrome, genetic damage of
Trichodermal syndrome mental retardation, genetic damage of
Trichodermodysplasia dental alterations, genetic damage of
Trichodysplasia xeroderma, genetic damage of
Trichoepithelioma multiple familial, genetic damage of
Trichofolliculloma, genetic damage of
Trichomalacia, genetic damage of
Trichomegaly cataract hereditary spherocytosis, genetic damage of
Trichomegaly retina pigmentary degeneration dwarfism, genetic damage of
Trichoodontoonychial dysplasia, genetic damage of
Trichopathophobia, genetic damage of (possible)
Trichorhinophalangeal syndrome type I, genetic damage of
Trichorhinophalangeal syndrome type II, genetic damage of
Trichorhinophalangeal syndrome type III, genetic damage of
Trichostasis spinulosa, genetic damage of
Trichothiodystrophy (generic term), genetic damage of
Trichothiodystrophy sun sensitivity, genetic damage of
Trichothiodystrophy with congenital ichtyosis, genetic damage of
Tricuspid atresia, genetic damage of
Tricuspid dysplasia, genetic damage of
Trigonocephaly bifid nose acral anomalies, genetic damage of
Trigonocephaly broad thumbs, genetic damage of
Trigonocephaly ptosis coloboma, genetic damage of
Trigonocephaly ptosis mental retardation, genetic damage of
Trigonomacrocephaly tibial defect polydactyly, genetic damage of
Trihydroxycholestanoylcoa oxidase isolated deficiency, genetic damage of

Trisomy 5q, genetic damage of
Trisomy 6, genetic damage of
Trisomy 6p, genetic damage of
Trisomy 6q, genetic damage of
Trisomy 7 mosaicism, genetic damage of
Trisomy 7p, genetic damage of
Trisomy 7p13 p12 2, genetic damage of
Trisomy 7q, genetic damage of
Trisomy 8, genetic damage of
Trisomy 8 Mosaicism, genetic damage of
Trisomy 8p, genetic damage of
Trisomy 8q, genetic damage of
Trisomy 9, genetic damage of
Trisomy 9 mosaic, genetic damage of
Trisomy 9 mosaicism, genetic damage of
Trisomy 9 translocation, genetic damage of
Trisomy 9p partial, genetic damage of
Trisomy 9q, genetic damage of
Trisomy 9q32, genetic damage of
Trisomy Partial 8, genetic damage of
Trisomy Xp3, genetic damage of
Trisomy Xpter Xq13, genetic damage of
Trisomy Xq, genetic damage of
Trisomy Xq25, genetic damage of
Trochlear dysplasia, genetic damage of
Trophoblastic neoplasms (gestational trophoblastic disease), genetic damage of
Trophoblastic tumor, genetic damage of
Tropical spastic paraparesis, genetic damage of
Tropophobia, genetic damage of (possible)
Troyer syndrome, genetic damage of
True hermaphroditism, genetic damage of
Trueb Burg Bottani syndrome, genetic damage of

Trypanophobia, genetic damage of (possible)
Tsao Ellingson syndrome, genetic damage of
Tsukahara Azuno Kajii syndrome, genetic damage of
Tsukahara Kajii syndrome, genetic damage of
Tsukuhara syndrome, genetic damage of
Tuberous sclerosis, genetic damage of
Tuberous sclerosis, type 1, genetic damage of
Tuberous sclerosis, type 2, genetic damage of
Tucker syndrome, genetic damage of
Tuffli Laxova syndrome, genetic damage of
Tufted angioma, genetic damage of (possible)
Tunglang Savage Bellman syndrome, genetic damage of
Turcot syndrome, genetic damage of
Turner Kieser syndrome, genetic damage of
Turner Morgani Albright, genetic damage of
Turner phenotype with normal karyotype, genetic damage of
Turner's syndrome, genetic damage of
Turner-like syndrome, genetic damage of
Tutuncuoglu syndrome, genetic damage of
Tyrosine transaminase deficiency, genetic damage of
Tyrosine-oxidase temporary deficiency, genetic damage of
Tyrosinemia, genetic damage of
Tyrosinemia type 1, genetic damage of
Tyrosinemia type 2, genetic damage of

U

Upton Young syndrome, genetic damage of
Urachal cancer, genetic damage of
Urachal cyst, genetic damage of
Urban Rogers Meyer syndrome, genetic damage of
Urban Schosser Spohn syndrome, genetic damage of
Urea cycle enzymopathies, genetic damage of
Uremia, genetic damage of
Urethral obstruction sequence, genetic damage of
Uridine monophosphate synthetase deficiency, genetic damage of
Urinary calculi, genetic damage of (possible)
Urinary tract neoplasm, genetic damage of
Urioste Martinez Frias syndrome, genetic damage of
Urogenital adysplasia, genetic damage of
Urophathy distal obstructive polydactyly, genetic damage of
Urticaria, genetic damage of
Urticaria pigmentosa, genetic damage of
Urticaria-deafness-amyloidosis, genetic damage of
Usher syndrome, genetic damage of
Usher syndrome, type 1A, genetic damage of
Usher syndrome, type 1B, genetic damage of
Usher syndrome, type 1C, genetic damage of
Usher syndrome, type 1D, genetic damage of
Usher syndrome, type 1E, genetic damage of
Usher syndrome, type 2A, genetic damage of
Usher syndrome, type 2B, genetic damage of
Usher syndrome, type 3, genetic damage of
Uveal diseases, genetic damage of
Uveitis, genetic damage of
Uveitis, anterior, genetic damage of
Uveitis, posterior, genetic damage of

V

VACTERL association, genetic damage of
Vacterl hydrocephaly, genetic damage of
Vacuolar myopathy, genetic damage of
Vagina, absence of, genetic damage of
Vagneur Triolle Ripert syndrome, genetic damage of
Valinemia, genetic damage of
Valvular dysplasia of the child, genetic damage of
Van Allen Myhre syndrome, genetic damage of
Van Bogaert disease, genetic damage of
Van Den Berghe Dequeker syndrome, genetic damage of
Van Den Bosch syndrome, genetic damage of
Van Den Ende Brunner syndrome, genetic damage of
Van der Woude syndrome, genetic damage of
Van Goethem syndrome, genetic damage of
Van Maldergem Wetzburger Verloes syndrome, genetic damage of
Van Regemorter Pierquin Vamos syndrome, genetic damage of
Varadi Papp syndrome, genetic damage of
Variegate porphyria, genetic damage of
Vas deferens, congenital bilateral aplasia of, genetic damage of
Vascular disruption sequence, genetic damage of
Vascular malformations of the brain, genetic damage of
Vascular malposition, genetic damage of
Vascular purpura, genetic damage of
Vasculitis hypersensitivity, genetic damage of
Vasculitis, cutaneous necrotizing, genetic damage of
Vasopressin-resistant diabetes insipidus, genetic damage of
Vasquez Hurst Sotos syndrome, genetic damage of
VATER association, genetic damage of
Vein of Galen aneurysm, genetic damage of
Velocardiofacial syndrome, genetic damage of
Velofacioskeletal syndrome, genetic damage of
Velopharyngeal incompetence, genetic damage of

Venencie Powell Winkelmann syndrome, genetic damage of
Ventricular extrasystoles perodactyly Robin sequence, genetic damage of
Ventricular familial preexcitation syndrome, genetic damage of
Ventricular fibrillation, idiopathic, genetic damage of
Ventricular septal defects, genetic damage of
Ventriculo-arterial discordance, isolated, genetic damage of
Ventruto Digirolamo Festa syndrome, genetic damage of
Venustraphobia, genetic damage of (possible)
Verbophobia, genetic damage of (possible)
Verloes Bourguignon syndrome, genetic damage of
Verloes David syndrome, genetic damage of
Verloes Gillerot Fryns syndrome, genetic damage of
Verloes Van Maldergem Marneffe syndrome, genetic damage of
Verloove Vanhorick Brubakk syndrome, genetic damage of
Verminiphobia, genetic damage of (possible)
Vernal keratoconjunctivitis, genetic damage of
Verneuil disease, genetic damage of
Verrucous nevus, genetic damage of
Verrucous nevus acanthokeratolytic, genetic damage of
Vertebral body fusion overgrowth, genetic damage of
Vertebral fusion posterior lumbosacral blepharoptosis, genetic damage of
Vertical talus, genetic damage of
Vestibulocochlear dysfunction progressive familial, genetic damage of
Vestiphobia, genetic damage of (possible)
Viljoen Kallis Voges syndrome, genetic damage of
Viljoen Smart syndrome, genetic damage of
Viljoen Winship syndrome, genetic damage of
Vipoma, genetic damage of
Virginitiphobia, genetic damage of (possible)
Virilism, genetic damage of

Virilizing ovarian tumor, genetic damage of

Virus associated hemophagocytic syndrome, genetic damage of (possible)

Visceral myopathy familial external ophthalmoplegia, genetic damage of

Viscero-atrial heterotaxia, genetic damage of

Vitamin B12 responsive methylmalonic acidemia, cbl A, genetic damage of

Vitamin B12 responsive methylmalonicaciduria, genetic damage of

Vitamin D resistant rickets, genetic damage of

Vitiligo, genetic damage of

Vitiligo mental retardation facial dysmorphism uremia, genetic damage of

Vitiligo psychomotor retardation cleft palate facial dysmorphism, genetic damage of

Vitreoretinal degeneration, genetic damage of

Vitreoretinochoroidopathy dominant, genetic damage of

VKH, genetic damage of

VLCAD deficiency, genetic damage of

Vocal cord dysfunction familial, genetic damage of

Vohwinkel syndrome, genetic damage of

Von Gierke disease, genetic damage of

Von Recklinghausen disease, genetic damage of

Von Voss Cherstvoy syndrome, genetic damage of

Von Willebrand disease, genetic damage of

Von Willebrand disease, dominant form, genetic damage of

Von Willebrand disease, recessive form, genetic damage of

Vulvar vestibulitis syndrome, genetic damage of

Vulvoldynia, genetic damage of (possible)

W

W syndrome, genetic damage of
Waaler Aarskog syndrome, genetic damage of
Waardenburg syndrome, genetic damage of
Waardenburg syndrome type 1, genetic damage of
Waardenburg syndrome type 2, genetic damage of
Waardenburg syndrome type 2A, genetic damage of
Waardenburg syndrome type 2B, genetic damage of
Waardenburg syndrome, type 3, genetic damage of
Waardenburg syndrome, type 4, genetic damage of
Waardenburg type Pierpont, genetic damage of
Waardenburg-Shah syndrome, genetic damage of
Wagner disease, genetic damage of
Wagner-Stickler syndrome, genetic damage of
WAGR syndrome, genetic damage of
Walbaum Titran Durieux Crepin syndrome, genetic damage of
Waldenstrom macroglobulinemia, genetic damage of
Waldmann disease, genetic damage of
Walker Dyson syndrome, genetic damage of
Wallerian degeneration, genetic damage of
Wallis Zieff Goldblatt syndrome, genetic damage of
Wandering spleen, genetic damage of
Warburg Sjo Fledelius syndrome, genetic damage of
Warburg Thomsen syndrome, genetic damage of
Warburton Anyane Yeboa syndrome, genetic damage of
Warkany, genetic damage of
Warm-reacting-antibody hemolytic anemia, genetic damage of
Warman Mulliken Hayward syndrome, genetic damage of
Watermelon stomach, genetic damage of
Watson Alagille syndrome, genetic damage of
Watson syndrome, genetic damage of
Weaver Johnson syndrome, genetic damage of
Weaver like syndrome, genetic damage of

Weaver syndrome, genetic damage of
Weaver Williams syndrome, genetic damage of
Weber Parkes syndrome, genetic damage of
Weber Sturge Dimitri syndrome, genetic damage of
Weber-Christian disease, genetic damage of
Webster Deming syndrome, genetic damage of
Wegener's granulomatosis, genetic damage of
Wegmann Jones Smith syndrome, genetic damage of
Weil syndrome, genetic damage of
Weill-Marchesani syndrome, genetic damage of
Weinstein Kliman Scully syndrome, genetic damage of
Weismann Netter Stuhl syndrome, genetic damage of
Weismann Netter syndrome, genetic damage of
Weissenbacher Zweymuller syndrome, genetic damage of
Welander distal myopathy, Swedish type, genetic damage of
Weleber Hecht Bigley syndrome, genetic damage of
Wellesley Carmen French syndrome, genetic damage of
Wells Jankovic syndrome, genetic damage of
Wells syndrome, genetic damage of
Werdnig-Hoffmann disease, genetic damage of
Werner's syndrome, genetic damage of
Wernicke Korsakoff syndrome, genetic damage of
West syndrome, genetic damage of
Westerhof Beemer Cormane syndrome, genetic damage of
Western equine encephalitis, genetic damage of
Westphall disease, genetic damage of
Whipple disease, genetic damage of
Whitaker syndrome, genetic damage of
White forelock with malformations, genetic damage of
White matter hypoplasia corpus callosum agenesia mental retardation, genetic damage of
Whyte Murphy syndrome, genetic damage of
Wiccaphobia, genetic damage of (possible)

Wieacker syndrome, genetic damage of
Wieacker-Wolff syndrome, genetic damage of
Wiedemann Grosse Dibbern syndrome, genetic damage of
Wiedemann Oldigs Oppermann syndrome, genetic damage of
Wiedemann Opitz syndrome, genetic damage of
Wiedemann Rautenstrauch syndrome, genetic damage of
Wildervanck syndrome, genetic damage of
Wilkes Stevenson syndrome, genetic damage of
Wilkie Taylor Scambler syndrome, genetic damage of
Willebrand disease type 1, genetic damage of
Willebrand disease type 2A, genetic damage of
Willebrand disease type 2B, genetic damage of
Willebrand disease type 2M, genetic damage of
Willebrand disease type 2N, genetic damage of
Willebrand disease type 3, genetic damage of
Willems de Vries syndrome, genetic damage of
Willi-Prader syndrome, genetic damage of
Williams syndrome, genetic damage of
Wilms tumor and pseudohermaphroditism, genetic damage of
Wilms tumor-aniridia, genetic damage of
Wilms tumour radial bilateral aplasia, genetic damage of
Wilms' tumor, genetic damage of
Wilson disease, genetic damage of
Wilson Turner syndrome, genetic damage of
Winchester syndrome, genetic damage of
Winkelman Bethge Pfeiffer syndrome, genetic damage of
Winship Viljoen Leary syndrome, genetic damage of
Winter Harding Hyde syndrome, genetic damage of
Winter Shortland Temple syndrome, genetic damage of
Wisconsin syndrome, genetic damage of
Wiskott Aldrich syndrome, genetic damage of
Witkop syndrome, genetic damage of
Wohlwill-Andrade syndrome, genetic damage of

Wolcott-Rallison syndrome, genetic damage of
Wolf-Hirschorn syndrome, genetic damage of
Wolff Zimmermann syndrome, genetic damage of
Wolff-Parkinson-White syndrome, genetic damage of
Wolfram syndrome, genetic damage of
Wolman disease, genetic damage of
Woodhouse Sakati syndrome, genetic damage of
Woods Black Norbury syndrome, genetic damage of
Woods Leversha Rogers syndrome, genetic damage of
Woolly hair hypotrichosis everted lower lip outstanding ears, genetic damage of
Woolly hair palmoplantar keratoderma cardiac anomalies, genetic damage of
Woolly hair, congenital, genetic damage of
Wooly hair syndrome, genetic damage of
Worster Drought syndrome, genetic damage of
Worth syndrome, genetic damage of
Wright Dick syndrome, genetic damage of
Wrinkly skin syndrome, genetic damage of
WT limb blood syndrome, genetic damage of
Wyburn-Mason's syndrome, genetic damage of

X

X fragile site folic acid type, genetic damage of

X linked juvenile retinoschisis, genetic damage of

X linked mental retardation Brooks type, genetic damage of

X linked mental retardation craniofacial abnormal microcephaly club, genetic damage of

X linked mental retardation de Silva type, genetic damage of

X linked mental retardation Hamel type, genetic damage of

X linked mental retardation short stature obesity, genetic damage of

X linked mental retardation type Gustavson, genetic damage of

X linked mental retardation type Martinez, genetic damage of

X linked mental retardation type Raynaud, genetic damage of

X linked mental retardation type Schutz, genetic damage of

X linked mental retardation type Snyder, genetic damage of

X linked mental retardation type Wittner, genetic damage of

X linked alpha thalassemia mental retardation syndrome (ATR-X), genetic damage of

X linked dominant, genetic damage of

X linked ichthyosis, genetic damage of

X linked juvenile retinoschisis, genetic damage of

X linked lymphoproliferative syndrome, genetic damage of

X linked mental retardation and macro-orchidism, genetic damage of

X linked mental retardation associated with marXq2, genetic damage of

X linked mental retardation-hypotonia, genetic damage of

X linked recessive, genetic damage of

X linked severe combined immunodeficiency, genetic damage of

Xanthic urolithiasis, genetic damage of

Xanthine oxydase deficiency, genetic damage of

Xanthinuria, genetic damage of

Xanthomatosis cerebrotendinous, genetic damage of

Y

Y chromosome deletions, genetic damage of
Yellow nail syndrome, genetic damage of (possible)
Yim Ebbin syndrome, genetic damage of
Yolk sac tumor, genetic damage of
Yorifuji Okuno syndrome, genetic damage of
Yoshimura-Takeshita syndrome, genetic damage of
Young Hugues syndrome, genetic damage of
Young Maders syndrome, genetic damage of
Young McKeever Squier syndrome, genetic damage of
Young Simpson syndrome, genetic damage of
Young syndrome, genetic damage of
Yunis Varon syndrome, genetic damage of

Z

About the Author

Nils K. Oeijord was born in Norway in 1947. A graduate of the Agricultural University of Norway, he also studied mathematics at the University of Trondheim, in Norway as well. He is a former assistant professor of mathematics at Tromsoe College, Norway, and is the author of several scientific works in Norwegian. He is currently a full time science writer. Nils K. Oeijord's first book in English, *Human Instincts Explained*, was published in 2000 (Vantage, New York). His second book in English, *A Dictionary of Human Instincts* (with Mitch C. Bronston), was published in 2001 (iUniverse), and his third book in English was *Human Behavior: The New Synthesis* (with Mitch C. Bronston) (iUniverse, 2001).

0-595-22565-9

www.ingramcontent.com/pod-product-compliance
Lightning Source LLC
Chambersburg PA
CBHW020738180526
45163CB00001B/276